AF787145

New Discoveries in Astronomical Data

New Discoveries in Astronomical Data

Guest Editors

Yanxia Zhang
A-Li Luo

MDPI

Basel • Beijing • Wuhan • Barcelona • Belgrade • Novi Sad • Cluj • Manchester

Guest Editors
Yanxia Zhang
National Astronomical Observatories
Chinese Academy of Sciences
Beijing
China

A-Li Luo
National Astronomical Observatories
Chinese Academy of Sciences
Beijing
China

Editorial Office
MDPI AG
Grosspeteranlage 5
4052 Basel, Switzerland

This is a reprint of the Special Issue, published open access by the journal *Universe* (ISSN 2218-1997), freely accessible at: https://www.mdpi.com/journal/universe/special_issues/UC90TJ0NUM.

For citation purposes, cite each article independently as indicated on the article page online and as indicated below:

Lastname, A.A.; Lastname, B.B. Article Title. *Journal Name* **Year**, *Volume Number*, Page Range.

ISBN 978-3-7258-5569-8 (Hbk)
ISBN 978-3-7258-5570-4 (PDF)
https://doi.org/10.3390/books978-3-7258-5570-4

© 2025 by the authors. Articles in this book are Open Access and distributed under the Creative Commons Attribution (CC BY) license. The book as a whole is distributed by MDPI under the terms and conditions of the Creative Commons Attribution-NonCommercial-NoDerivs (CC BY-NC-ND) license (https://creativecommons.org/licenses/by-nc-nd/4.0/).

Contents

About the Editors

Yanxia Zhang

Yanxia Zhang is a Professor at the National Astronomical Observatories, Chinese Academy of Sciences. She also serves as an Organizing Committee Member of IAU Commission B3 on Astroinformatics and Astrostatistics, an Associate Editor for Astrostatistics at *Frontiers in Astronomy and Space Sciences*, and a member of the editorial board of *Universe*. Her research interests include astrostatistics and astroinformatics, the application of artificial intelligence in astronomy, and the classification and identification of celestial objects, as well as the measurement of their physical parameters. She was awarded the International Astrostatistics Association (IAA) Fellow Prize in 2016.

A-Li Luo

A-Li Luo is a Professor and the Director of the LAMOST Operation and Development Center at the National Astronomical Observatories, Chinese Academy of Sciences (NAOC). He is also the head of the AI Working Group at NAOC, the Vice Chair of the Commission on Astronomical Technology of the Chinese Astronomical Society, and a member of the Committee on Data and Literature at the International Astronomical Union (IAU). His research expertise includes massive spectroscopic data processing, astronomical data mining and analysis, and the application of artificial intelligence in astronomy. He has received several prestigious awards, including the Outstanding Science and Technology Achievement Prize from the Chinese Academy of Sciences (Collective, 2019) and the Second Prize of the Beijing Science and Technology Progress Award (First Completer, 2020).

universe MDPI

Editorial

Editorial for the Special Issue "New Discoveries in Astronomical Data"

Yanxia Zhang [1,*] and A-Li Luo [1,2]

1 CAS Key Laboratory of Optical Astronomy, National Astronomical Observatories, Chinese Academy of Sciences, Beijing 100101, China
2 University of Chinese Academy of Sciences, Beijing 100049, China
* Correspondence: zyx@bao.ac.cn

Over the past decade, astronomy has shifted from a data-starved to a data-drenched science [1,2]. The Vera C. Rubin Observatory's Legacy Survey of Space and Time (LSST), ESA's Euclid, and the Square Kilometre Array (SKA) pathfinders are already delivering terabytes of images, spectra, and time-series each night. Traditional analysis pipelines—calibrated for megabyte-scale data sets—cannot keep pace, and a new generation of artificial intelligence (AI) and machine learning (ML) techniques has become indispensable [3–6]. Yet the very power of these methods has revealed a critical gap: the "black-box" nature of many state-of-the-art algorithms threatens the reproducibility, interpretability, and ultimately the credibility of the discoveries they enable.

This Special Issue was conceived to bridge three intertwined gaps:

Trustworthy AI. How do we exploit deep neural networks, self-supervised learning, and real-time inference without sacrificing the transparency required by the scientific method?

Domain-specific bottlenecks. Solar radio spectroscopy, wide-field meteor monitoring, CO clump cataloguing, and RR Lyrae parameterisation each suffer from small training sets, strong selection effects, or instrument-specific artefacts that generic ML toolkits ignore.

Environmental integrity. As facilities push into previously "quiet" frequency bands, radio-frequency interference (RFI) threatens both current and future observations; rigorous site-protection modelling is now mission-critical.

The eight articles collected here address these challenges head-on and chart a coherent path forward.

Lieu's comprehensive review opens the volume by framing the interpretability crisis [Contribution 1]. She distils the vocabulary—transparency, interpretability, explainability—and showcases astronomy-specific tools (symbolic regression, physics-informed neural networks, and SHAP/LIME visualizations) that convert predictive power into physical insight. Her call for interdisciplinary benchmarks sets the tone for every contribution that follows.

The next contribution, on solar physics, demonstrates the power of self-supervised learning when labeled examples are scarce [Contribution 2]. By adapting BERT-style masking to Vision Transformers, the authors achieve 99.5% accuracy in classifying solar radio bursts—outperforming supervised baselines—while revealing that a 75% masking rate (far above NLP standards) is optimal for redundant spectral images. The study establishes a new transfer-learning paradigm for small-sample astronomical data.

Accurate environmental protection starts with accurate propagation models [Contribution 3]. The RFI analysis software presented here integrates global terrain, meteorology, and satellite ephemerides into an ITU-R-compliant framework that has already guided site-selection for the Five-hundred-meter Aperture Spherical Telescope (FAST) and the planned Qitai Telescope (QTT) [Contribution 4]. Freely available binaries and an interactive map-based GUI invite the community to extend these protections to next-generation arrays.

Single-site meteor detection is notoriously plagued by false positives from aircraft, satellites, and sensor noise. GWAC's new algorithm exploits temporal differencing, probabilistic Hough transforms, and light-curve morphology to reach 90% accuracy on 4K $\times$ 4K frames sampled every 15 s—a three-fold speed-up over prior work [Contribution 5]. The open-source pipeline is already ingesting real-time data from the Geminid stream and will inform the forthcoming Chinese Space Station Telescope transient key project.

Cataloguing molecular clouds has traditionally required human-tuned thresholds and dozens of free parameters [Contribution 6]. MCD-YOLOv5 replaces this tedium with a two-stage deep-learning workflow: an attention-augmented YOLOv5 detector slices Galactic plane images, while Density Peak Clustering stitches velocity channels into 3D clumps. Trained on 10,000 synthetic but realistic MWISP cubes, the network attains 98% recall with only two tunable parameters—an order-of-magnitude reduction compared with FellWalker or ClumpFind.

The RR Lyrae study shows how neural emulators can invert light-curve morphology into fundamental stellar parameters [Contribution 7]. A dense hydrodynamic grid and a four-layer ANN recover mass, luminosity, effective temperature, and metallicity from TESS single-band photometry alone. The derived period–luminosity–metallicity relation is consistent with theoretical predictions and demonstrates a fast, scalable route to precision stellar astrophysics for LSST's multi-band light curves.

Finally, the study on H II regions in NGC 2403 concludes that these regions are primarily star-forming with a clear inside-out evolution pattern [Contribution 8]. The N_2O_2 diagnostic is identified as the most reliable for estimating metallicity due to its insensitivity to the ionization parameter and age. The study highlights limitations in reproducing certain emission lines, particularly [O II], suggesting future work should consider the effects of stellar rotation or binary populations to improve model accuracy. This research provides valuable insights into the properties and evolution of H II regions in NGC 2403, contributing to the broader understanding of star formation and galaxy evolution.

Taken together, these works deliver a clear message: **interpretable, domain-aware AI is no longer optional; it is the prerequisite for turning tomorrow's exabyte surveys into lasting scientific knowledge**.

Looking forward, five priorities emerge:

Mechanistic interpretability. We must move from post hoc saliency maps to architectures that embed conservation laws, radiative transfer, and gravitational dynamics directly into their loss functions.

Causal inference. Correlations uncovered by neural networks need to be interrogated with counterfactual simulations to disentangle astrophysical causation from observational bias.

Scalable, uncertainty-aware pipelines. As data volumes grow, even lightweight explainability tools become computationally prohibitive. Hardware-aware pruning and probabilistic neural networks will be essential.

Cross-messenger consistency. Joint electromagnetic, gravitational-wave, and neutrino data sets require federated learning frameworks that respect proprietary formats while preserving interpretability.

Ethical stewardship of the spectrum. RFI mitigation strategies must evolve in tandem with satellite mega-constellations; open-source tools like those presented here should become community standards endorsed by the IAU and national regulators.

We invite the readership to treat this Special Issue not as a collection of isolated advances but as a blueprint for the next decade of trustworthy, data-driven discovery.

The code, models, and data sets accompanying each paper are released under permissive licences precisely to encourage rapid iteration and broader adoption. Let us build on these foundations—together—to ensure that the coming avalanche of astronomical data yields insights that are as reliable and profound as the night sky itself.

As AI technology continues to evolve, its applications in astronomy are expected to expand and deepen, leading to even more groundbreaking discoveries [7–14]. The future of AI in astronomy holds several promising directions:

Advanced Machine Learning Models: The development of more sophisticated machine learning models will likely uncover new types of phenomena and optimize space missions in ways not yet imagined. These models will be capable of handling the increasing volume and complexity of astronomical data, leading to more accurate predictions and discoveries.

Data Fusion and Multi-Messenger Astronomy: The ability to effectively fuse data from multiple sources and instruments will become increasingly important. AI will play a crucial role in integrating and analyzing data from different telescopes, observatories, and missions, enabling a more comprehensive understanding of cosmic events. This will be particularly important for multi-messenger astronomy, where data from gravitational waves, electromagnetic radiation, and other sources are combined to study cosmic phenomena.

Enhanced Observational Capabilities: AI will continue to improve the capabilities of ground-based and space-based telescopes. This includes advancements in adaptive optics, which can correct for atmospheric distortions in real-time, and the development of new algorithms for image processing and data analysis. The integration of AI with upcoming observatories such as the SKA and the Rubin Observatory will enable unprecedented levels of detail and accuracy in astronomical observations.

Citizen Science and Public Engagement: AI can facilitate citizen science projects by automating tasks that would otherwise require significant human effort. This will allow more people to participate in astronomical research, contributing to the discovery and analysis of celestial objects. Additionally, AI-generated visualizations and educational tools will enhance public engagement and understanding of astronomy.

Addressing Challenges: While AI offers numerous benefits, it also presents challenges such as data quality, model interpretability, and computational resource requirements. Future research will focus on developing robust data quality enhancement algorithms, creating interpretable AI models, and optimizing computational resources to ensure the efficient and effective use of AI in astronomy.

In conclusion, AI has already transformed the field of astronomy by enhancing data processing, enabling new discoveries, and improving observational capabilities. As technology continues to advance, AI will play an increasingly vital role in unravelling the mysteries of the universe, making astronomy more accessible and engaging for both researchers and the public.

Author Contributions: Conceptualization, Y.Z. and A.-L.L.; writing—original draft preparation, Y.Z.; writing—review and editing, Y.Z. and A.-L.L.; funding acquisition, Y.Z. and A.-L.L. All authors have read and agreed to the published version of the manuscript.

Funding: This research was funded by the National Natural Science Foundation of China (grant Nos. 12273076, 12273075 and 12133001).

Conflicts of Interest: The authors declare no conflicts of interest.

List of Contributions:

1. Lieu, M. A Comprehensive Guide to Interpretable AI-Powered Discoveries in Astronomy. *Universe* **2025**, *11*, 187. https://doi.org/10.3390/universe11060187.
2. Li, S.; Yuan, G.; Chen, J.; Tan, C.; Zhou, H. Self-Supervised Learning for Solar Radio Spectrum Classification. *Universe* **2022**, *8*, 656. https://doi.org/10.3390/universe8120656.
3. Wang, J.; Zhao, Y.; Yang, C.; Shi, Y.; Hao, Y.; Zhang, H.; Sun, J.; Luo, D. The Analysis and Verification of IMT-2000 Base Station Interference Characteristics in the FAST Radio Quiet Zone. *Universe* **2023**, *9*, 248. https://doi.org/10.3390/universe9060248.
4. Wang, Y.; Zhang, H.; Wang, J.; Huang, S.; Hu, H.; Yang, C. A Software for RFI Analysis of Radio Environment around Radio Telescope. *Universe* **2023**, *9*, 277. https://doi.org/10.3390/universe9060277.
5. Chen, Y.; Li, G.; Liu, C.; Qiu, B.; Shan, Q.; Li, M. A Meteor Detection Algorithm for GWAC System. *Universe* **2023**, *9*, 468. https://doi.org/10.3390/universe9110468.
6. Hu, J.-B.; Huang, Y.; Zheng, S.; Chen, Z.-W.; Zeng, X.-Y.; Luo, X.-Y.; Long, C. Molecular-Clump Detection Based on an Improved YOLOv5 Joint Density Peak Clustering. *Universe* **2023**, *9*, 480. https://doi.org/10.3390/universe9110480.
7. Kumar, N.; Singh, H.P.; Malkov, O.; Joshi, S.; Tan, K.; Prugniel, P.; Bhardwaj, A. Extraction of Physical Parameters of RRab Variables Using Neural Network Based Interpolator. *Universe* **2025**, *11*, 207. https://doi.org/10.3390/universe11070207.
8. Wu, Q.-M.; Mao, Y.-W.; Lin, L.; Zou, H.; Wang, S.-T. Spectroscopic Observations and Emission-Line Diagnoses for H ii Regions in the Late-Type Spiral Galaxy NGC 2403. *Universe* **2025**, *11*, 280. https://doi.org/10.3390/universe11080280.

References

1. Khalil, M.; Said, M.; Osman, H.; Ahmed, B.; Ahmed, D.; Younis, N.; Maher, B.; Osama, M.; Ashmawy, M. Big data in astronomy: From evolution to revolution. *Int. J. Adv. Astron.* **2019**, *7*, 11. [CrossRef]
2. Mickaelian, A.M. Big Data in Astronomy: Surveys, Catalogs, Databases and Archives. *Commun. BAO* **2020**, *67*, 159. [CrossRef]
3. Wang, K.; Guo, P.; Yu, F.; Duan, L.; Wang, Y.; Du, H. Computational Intelligence in Astronomy: A Survey. *Int. J. Comput. Intell. Syst.* **2018**, *11*, 575. [CrossRef]
4. Das, S.; Dey, A.; Roy, N. Applications of Artificial Intelligence in Machine Learning: Review and Prospect. *Int. J. Comput. Appl.* **2015**, *115*, 9. [CrossRef]
5. Meher, S.K.; Panda, G. Deep learning in astronomy: A tutorial perspective. *Eur. Phys. J. Spec. Top.* **2021**, *230*, 2285. [CrossRef]
6. Thatikonda, V.K.; Golla, Y.; Mudunuri, H.R. Developments in Artificial Intelligence and Machine Learning: Recent Advances and Future Prospects. *ESP Int. J. Adv. Sci. Technol.* **2024**, *2*, 1–5.
7. Djorgovski, S.G.; Mahabal, A.A.; Graham, M.J.; Polsterer, K.; Krone-Martins, A. Applications of AI in Astronomy. *arXiv* **2022**, arXiv:2212.01493. [CrossRef]
8. Fluke, C.J.; Jacobs, C. Surveying the reach and maturity of machine learning and artificial intelligence in astronomy. *Wiley Interdiscip. Rev. Data Min. Knowl. Discov.* **2020**, *10*, e1349. [CrossRef]
9. Impey, C. How AI Is Helping Astronomers. EarthSky. Available online: https://earthsky.org/space/artifical-intelligence-ai-is-helping-astronomers-make-new-discoveries/ (accessed on 23 May 2023).
10. Pruthi, M.N. Artificial Intelligence in Astronomy. *Int. J. Res. Appl. Sci. Eng. Technol. (IJRASET)* **2019**, *7*, 903. [CrossRef]
11. Huang, K.; Hu, T.; Cai, J.; Pan, X.; Hou, Y.; Xu, L.; Wang, H.; Zhang, Y.; Cui, X. Artificial Intelligence in Astronomical Optical Telescopes: Present Status and Future Perspectives. *Universe* **2024**, *10*, 210. [CrossRef]
12. Smith, M.J.; Geach, J.E. Astronomia ex machina: A history, primer and outlook on neural networks in astronomy. *R. Soc. Open Sci.* **2023**, *10*, 221454. [CrossRef] [PubMed]
13. Sharma, P.; Vaidya, B.; Wadadekar, Y.; Bagla, J.; Chatterjee, P.; Hanasoge, S.; Kumar, P.; Mukherjee, D.; Philip, N.S.; Singh, N. Computational Astrophysics, Data Science & AI/ML in Astronomy: A Perspective from Indian Community. *arXiv* **2025**, arXiv:2501.03876. [CrossRef]
14. Znamenskiy, V. Artificial Intelligence and Astronomy: Lab Manual for Generative AI-Based Learning Activities. Available online: https://zenodo.org/records/15555802 (accessed on 28 August 2025).

Disclaimer/Publisher's Note: The statements, opinions and data contained in all publications are solely those of the individual author(s) and contributor(s) and not of MDPI and/or the editor(s). MDPI and/or the editor(s) disclaim responsibility for any injury to people or property resulting from any ideas, methods, instructions or products referred to in the content.

universe

MDPI

Article

Self-Supervised Learning for Solar Radio Spectrum Classification

Siqi Li [1], Guowu Yuan [1,2,*], Jian Chen [1], Chengming Tan [3,4] and Hao Zhou [1]

1 School of Information Science & Engineering, Yunnan University, Kunming 650500, China
2 CAS Key Laboratory of Solar Activity, National Astronomical Observatories, Beijing 100012, China
3 State Key Laboratory of Space Weather, National Space Science Center, Beijing 100864, China
4 School of Astronomy and Space Sciences, University of Chinese Academy of Sciences, Beijing 100049, China
* Correspondence: gwyuan@ynu.edu.cn

Abstract: Solar radio observation is an important way to study the Sun. Solar radio bursts contain important information about solar activity. Therefore, real-time automatic detection and classification of solar radio bursts are of great value for subsequent solar physics research and space weather warnings. Traditional image classification methods based on deep learning often require considerable training data. To address insufficient solar radio spectrum images, transfer learning is generally used. However, the large difference between natural images and solar spectrum images has a large impact on the transfer learning effect. In this paper, we propose a self-supervised learning method for solar radio spectrum classification. Our method uses self-supervised training with a self-masking approach in natural language processing. Self-supervised learning is more conducive to learning the essential information about images compared with supervised methods, and it is more suitable for transfer learning. First, the method pre-trains using a large amount of other existing data. Then, the trained model is fine-tuned on the solar radio spectrum dataset. Experiments show that the method achieves a classification accuracy similar to that of convolutional neural networks and Transformer networks with supervised training.

Keywords: solar radio spectrum; deep learning; self-supervised learning; transfer learning

1. Introduction

Solar radio spectrum observations are an important tool for studying solar outbursts, which contain important information about solar activity [1]. The solar radio spectra are divided into various types, corresponding to different physical events [2,3]. With the development of radio spectrometers and the massive observational data trend, manual detection and classification of solar radio spectra can no longer meet the needs of research. Therefore, it is important to automatically detect and classify solar radio bursts from this massive information efficiently and rapidly for subsequent scientific research and space weather warning and forecasting.

The solar broadband radio spectrometers (SBRS) at the National Astronomical Observatory of the Chinese Academy of Sciences were put into operation during the 23rd solar activity cycle [4]. The devices have produced a large number of observations. However, since solar radio bursts are a low-probability event, the observed spectra of solar radio bursts are very small. Additionally, due to the presence of interference, the raw data are not clearly characterized and it is difficult to quickly distinguish between different kinds of data. This creates difficulties for subsequent astronomical studies [5]. Therefore, it is of great help for solar physics research to classify solar radio spectra accurately, quickly and automatically.

With the rapid development of hardware levels and artificial intelligence, an increasing number of deep learning models and algorithms are used to solve tasks related to natural language processing [6] and computer vision [7,8]. For astronomy problems, many previous works were carried out on solar radio spectrum classification and these works have used image processing algorithms, neural network models and deep learning [9]. P. J. Zhang et al. designed an event recognition analysis system that can automatically detect solar type III radio bursts. This system used Hough transform to recognize the line segment associated with type III bursts in the dynamic spectra [10]. However, the computational parameters of this method must be artificially designed and are not universally applicable. In recent years, with the development of convolutional neural networks (CNN) [11,12], long short-term memory (LSTM) networks [13], and deep confidence networks [14], many of these methods were applied to the classification of solar radio spectra. S. M. J. Jalali introduced LSTM [15], which was combined with a CNN to propose the CNN–LSTM approach [16]. This method improves performance with similar time consumption. B. Yan used a feature pyramid network (FPN) as a backbone network [17] and used ResNet to extract features [18]. By simply connecting to the structure, FPN fuses features of different scales and different levels of semantics. The performance of detection is improved without affecting the speed of detection. In other related studies, a classification algorithm based on joint convolutional neural networks and transfer learning was proposed by using the inherent correlation between natural datasets and astronomical datasets. In addition, a cost-sensitive multiclassification loss function was proposed to make the network pay more attention to categories with fewer samples during the training process and use the meta-learning method to classify the solar radio spectrum with fewer samples. In addition, H. Salmane et al. proposed automatic identification methods for specific types of solar radio burst structures [19].

In current research on image classification, a large amount of training data is generally needed. However, since solar radio bursts are low-probability events, there are few samples. To solve the problem of fewer data, some studies have used transfer learning technology and improved loss functions. However, the large difference between natural images and solar radio spectrum images is not conducive to the application of transfer learning.

In natural language processing, self-supervised learning methods have become popular and these methods no longer require large amounts of labeled data in the training step. Researchers can design some rules to let the data supervise their own training [20]. For example, bidirectional encoder representations from Transformer (BERT) are designed as a kind of fill-in-the-blank method by masking some words in a sentence and then letting the network guess those words. In natural language processing, self-supervised learning is usually based on autoregressive language modeling in generative pre-training (GPT) and mask self-coding in BERT. Their basic principle is to delete some data and let the network learn to predict the deleted content.

In this paper, a solar radio spectrum classification method based on self-supervised learning is proposed. The method uses BERT to train the network to classify solar radio spectrum images by randomly masking a portion of the solar radio spectrum images and letting the network fill in the blank. The main contribution of this paper is that we apply self-supervised learning to classify the solar radio spectrum for the first time. This method is more conducive to using transfer learning. This paper can provide a reference for other small sample data classifications in astronomy.

2. Solar Radio Spectrum Dataset and Its Preprocessing

The solar radio spectra are obtained from the solar broadband radio spectrometer (SBRS) at the National Astronomical Observatory of the Chinese Academy of Sciences. The raw data are stored in binary format and visualized to obtain the solar radio spectrum images. The vertical axis of the solar radio spectrum image represents the frequency of the spectrum, the horizontal axis represents the time, and each pixel value represents the radio flux of the Sun at a certain time and frequency. When displayed as a grayscale image, white indicates high solar radio flux, black indicates low solar radio flux, and the whole image represents the solar radio flux at a frequency over a period of time. The solar radio spectra are divided into three categories: burst, calibration and non-burst, as shown in Figure 1. The classification tasks in this paper are aimed at these three categories.

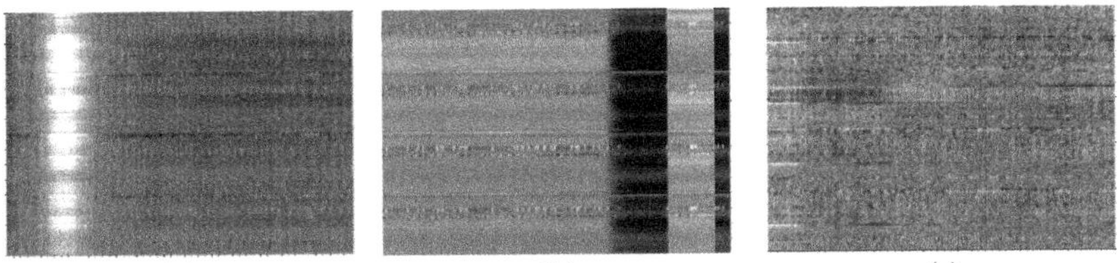

(a) (b) (c)

Figure 1. Solar radio spectrum data types. (**a**) Radio burst; (**b**) calibration; (**c**) non-burst.

The solar radio spectra images show that the original image is noisy and most of the noise is transverse stripe noise. This noise will affect the subsequent classification accuracy. The frequency channel normalization can reduce transverse stripe noise and make its features obvious, and the calculation method is as follows:

$$p'(x,y) = p(x,y) - \frac{1}{n}\sum_{y=0}^{n} p(x,y) + \frac{1}{mn}\sum_{x=0}^{m}\sum_{y=0}^{n} p(x,y) \quad (1)$$

where $p(x,y)$ represents the radio intensity at time x and frequency y on the spectrum and $p'(x,y)$ is the radio intensity after channel normalization. The final result is shown in Figure 2. The noise of the horizontal stripe is significantly reduced and the burst is more obvious. It makes the features of the image more obvious and facilitates the learning of the subsequent network.

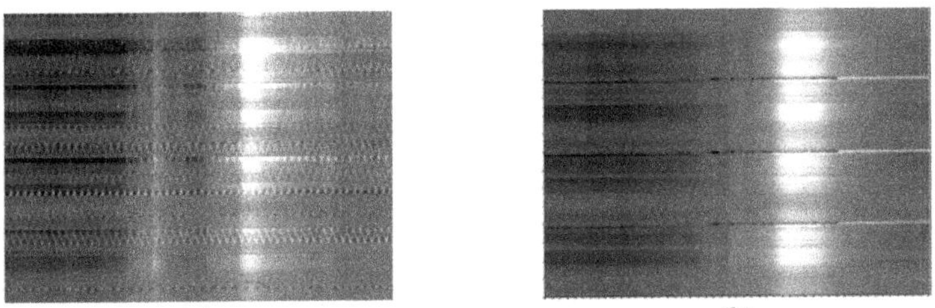

(a) (b)

Figure 2. Results of channel normalization. (**a**) Before channel normalization; (**b**) after channel normalization.

3. Method Section

3.1. Self-Supervised Learning

Self-supervised learning (SSL) is the main method used with the Transformer model to learn from large-scale unlabeled datasets [21]. The basic idea of SSL is to fill in the gaps. It masks or hides some parts of the input and uses observable parts to predict hidden parts [22]. Another effective method of self-supervised learning is contrastive learning [23]. In this case, we usually learn the feature representation of samples by creating positive and negative samples and comparing the data with positive samples and negative samples in the feature space. The advantage of contrastive learning is that it does not need to reconstruct pixel-level details to obtain image features but only needs to learn differentiation in the feature space. However, the construction of positive and negative samples is a difficult point in contrastive learning.

SSL provides a promising learning paradigm since it enables learning from a vast amount of readily available nonannotated data. SSL is implemented in two steps. First, a model is trained to learn a meaningful representation of the underlying data by solving a pretext task. The pseudo labels for the pretext task are automatically generated based on data attributes and task definition. Therefore, a critical choice in SSL is the definition of a pretext task. Second, the model trained in the first step is fine-tuned on the downstream tasks using labeled data. The downstream tasks include image classification and object detection [24,25].

3.2. Self-Supervised Learning with Self-Masking

Self-supervised learning methods are widely used in natural language processing. However, due to the large difference between the CNN model in computer vision and the Transformer used in natural language processing, the natural language processing method cannot be effectively transferred to computer vision tasks. However, when Vision Transformer (ViT) was proposed [26], the channel between computer vision and natural language was opened [27–30].

We refer to the mask method used in the BERT and GPT in natural language and let the model learn to restore sentences after covering some words. When the methods are transferred to solar radio spectrum classification, the solar radio spectrum image is first divided into blocks, which are equivalent to a word in a sentence. Then, these blocks are masked randomly and sent to the network. The network is composed of the encoder and decoder of ViT. After the image is restored through model learning, the trained encoder is removed and connected to the fully connected layer for classification. The structure of the self-mask model is shown in Figure 3.

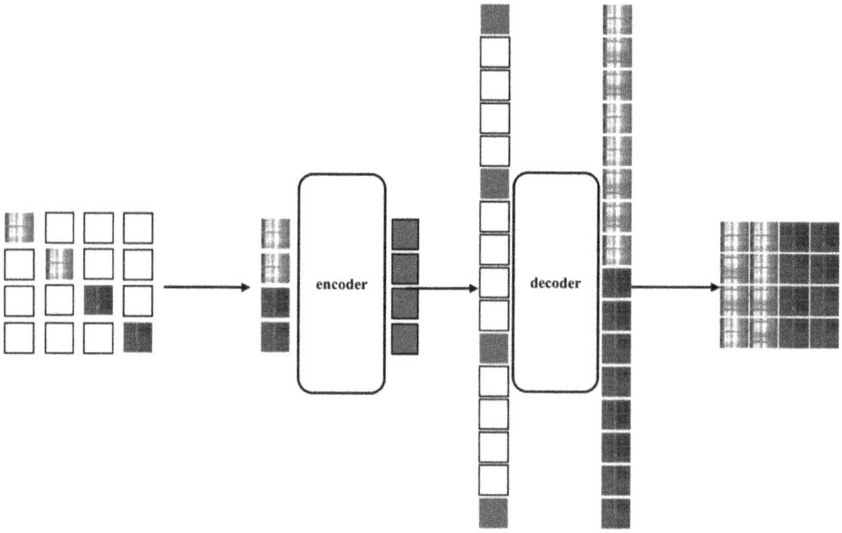

Figure 3. Self-masking model structure.

Compared with text, we can cover more parts due to the more redundant information in the image. This not only saves space and improves the model speed but also helps the model learn more information from solar radio spectrum images.

3.3. Encoder and Decoder

Our encoder is only derived from the ViT structure for visible unmasked blocks. Similarly to standard ViT, our encoder embeds patches by adding a linear projection of the positional embedding and then processes the dataset through a series of Transformer blocks. However, our encoder only needs to work on a small subset (e.g., 25%), which allows us to train very large encoders that require only a fraction of the computations and memory. The complete dataset is processed by a lightweight decoder.

The input to the decoder is a complete token set consisting of encoded visible blocks and mask tokens. Each mask token is a shared and learned vector that represents the missing blocks to be predicted. We add a position to all tokens in this complete set. If we do not do this, mask tokens will not have information about their positions in the image. The decoder also has another series of Transformer blocks.

The decoder is only used to perform image reconstruction tasks prior to training. Thus, the decoder architecture can be designed in a flexible way, independent of the encoder design. We experiment with very small decoders that are narrower and shallower than the encoder. For example, our default decoder is 10% smaller in computations per token than the encoder. In this asymmetric design, the complete set of tokens is processed by the lightweight decoder only, which significantly reduces the pre-training time. The encoder and decoder structures are shown in Figure 4.

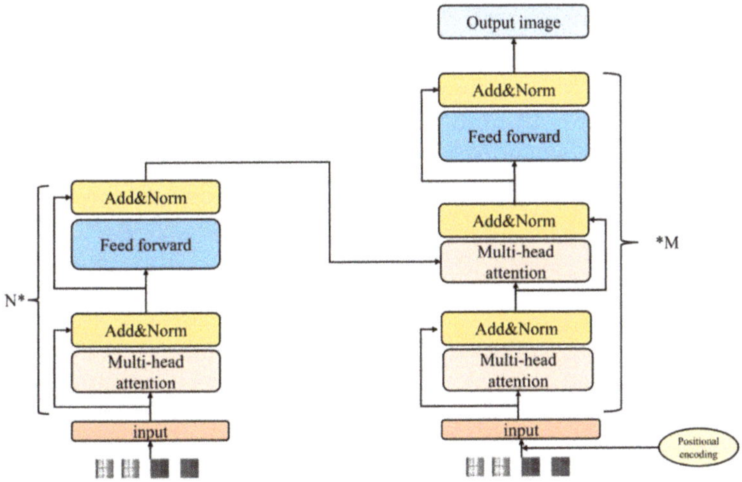

Figure 4. Encoder (**left**) and decoder (**right**) structures.

3.4. Data Enhancement

Because solar bursts are low-probability events, the solar radio burst spectra observed are not much. To improve the generalization ability and avoid the overfitting of the model, we augment the spectrum of solar radio bursts. We adopt a simple and data-independent augmentation method called mixup. Mixup can be implemented in only a few lines of code. A virtual training sample is constructed with minimal computational overhead. In an extensive evaluation, the results show that mixup improves the generalization error of the most advanced models in ImageNet, CIFAR, voice and tabular datasets. In addition, mixup helps to eliminate the memory of false labels, the sensitivity to confrontation samples and the instability of confrontation training. The formula for mixup data enhancement is as follows:

$$\widetilde{x} = \lambda x_i + (1 - \lambda) x_j \tag{2}$$

$$\widetilde{y} = \lambda y_i + (1 - \lambda) y_j \tag{3}$$

where x_i and x_j are two images randomly selected from the training set and y_i and y_j are their corresponding one-hot tags, respectively. Prior knowledge indicates that the linear interpolation of the sample images and the linear interpolation of the corresponding one-hot labels correspond. Mixup constructs a new sample and its one-hot label $\widetilde{x}, \widetilde{y}$ based on this prior knowledge. Among them, λ is obtained by the data distribution $\beta(\alpha, \alpha)$, and α is a super parameter.

By adjusting the super parameter α, we can adjust the proportion of interpolation between images. The research also shows that there is no good method to set α at present and the sensitivity to α is different in different datasets.

To further improve the generalization ability of the model and avoid overfitting, dropout is also introduced in this model. In deep learning, if the amount of data is small and the model is complex, the trained model will easily overfit. Dropout can alleviate the overfitting problem and achieve a regularization effect to a certain extent.

4. Experimental Dataset and Experimental Configuration

4.1. Experimental Dataset

A solar radio burst corresponds to a certain solar activity event, which is a low-probability event. The calibration data are relatively small and a large number of solar radio spectra are no-burst spectra. Therefore, there is an imbalance between the three types of samples.

Since the solar radio spectrum has two parts, left-handed and right-handed polarization, we separated the two parts so that the burst and calibration data can be expanded. The amount of solar radio data after amplification is shown in Table 1. The quantity of the three types of spectra is roughly in a balanced state, which alleviates the data imbalance problem. The experiment shows that the separation of these two parts has no effect on the results of solar radio spectrum classification.

Table 1. Number of solar radio spectrum samples.

Type	Training Set	Testing Set	Total
Non-burst	1648	412	2060
Burst	1476	369	1845
Calibration	1292	322	1614

After data enhancement, the total number of samples is 5519. We divided these randomly according to a proportion of approximately 8:2; 4415 labeled images were used as the training set and the remaining 1104 were used as the testing set. The specific number of samples in each category of the dataset is shown in Table 1.

4.2. Experimental Configuration and Evaluation Index

The software environment for our experiments is Windows 10, the programming platform is PyCharm, and the architecture is PyTorch. In the hardware device, the CPU is an Intel Core i7-10700k, the memory is 32 GB, and the GPU is a NVidia GeForce RTX 2080Ti. For pre-training, the batch size is 16, the image size is 224×224, the epoch is set to 300, the learning rate is 1.5×10^{-4}, the warmup learning rate is 10^{-6}, the warmup is 30 epochs, and the weight decay is set to 0.05. In the fine-tuning stage, the epoch is set to 50, the learning rate is 10^{-3}, the warmup learning rate is 10^{-6}, the warmup is five epochs, and the other parameters are unchanged.

In practice, the burst class has greater research value and greater impact on daily life, so we mainly focus on the burst class. The burst class is defined as a positive class, and the other two classes are defined as negative classes. We define TP as the number of samples that are positive and correctly classified as positive, FP as the number of samples that are negative but wrongly classified as positive, TN as the number of samples that are negative and correctly classified as negative, and FN as the number of samples that are positive but classified as negative. The evaluation metrics used in the experiment are accuracy, precision, recall, specificity, and *F*-score.

Accuracy refers to the proportion of the number of accurate samples classified by all categories to the total number of samples. It is calculated as follows:

$$Accuracy = \frac{TP + TN}{TP + TN + FP + FN} \tag{4}$$

Precision refers to the proportion of the number of correctly predicted positive samples to the total number of predicted positive samples. It is calculated as follows:

$$Precision = \frac{TP}{TP + FP} \tag{5}$$

Recall refers to the proportion of correctly predicted positive samples relative to all actual positive samples. It is calculated as follows:

$$Recall = \frac{TP}{TP + FN} \tag{6}$$

Specificity refers to the proportion of correctly predicted negative cases relative to all actual negative cases. It is calculated as follows:

$$Specificity = \frac{TN}{TN + FP} \tag{7}$$

The balanced *F*-score is used for the overall evaluation of precision and recall. It is calculated as follows:

$$F_1 = 2 \times \frac{Precision \times Recall}{Precision + Recall} \tag{8}$$

5. Experimental Results and Discussion

5.1. Effect of Masking Rate on Classification Accuracy

When using the self-masking model structure for training, we divide the solar radio spectrum image into blocks and then mask these blocks randomly and send them to the network for training. We found that different masking rates affect training and classification accuracy. After using different masking rates for the experiments, the influence of masking rates on classification accuracy is shown in Figure 5.

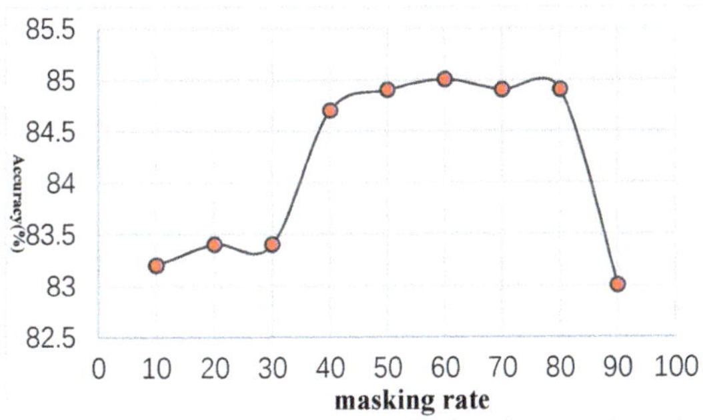

Figure 5. Influence of masking rate on classification accuracy.

In Figure 5, the experimental data show that a high masking rate corresponds to a high accuracy. When the masking rate is between 60% and 75%, the classification accuracy is the highest. The experimental results are counterintuitive. In natural language processing, the best masking rate of the BERT method is about 15%. In our results, the optimal masking rate is much higher. This may be because compared with text, images have more redundant information, while text contains more dense information.

Therefore, masking more information in the image can remove much redundant information unrelated to classification. This is more conducive to learning useful features and more essential information from the image, which is conducive to subsequent transfer learning. In this experiment, a random masking rate of 75% is finally selected and the masking effect is shown in Figure 6.

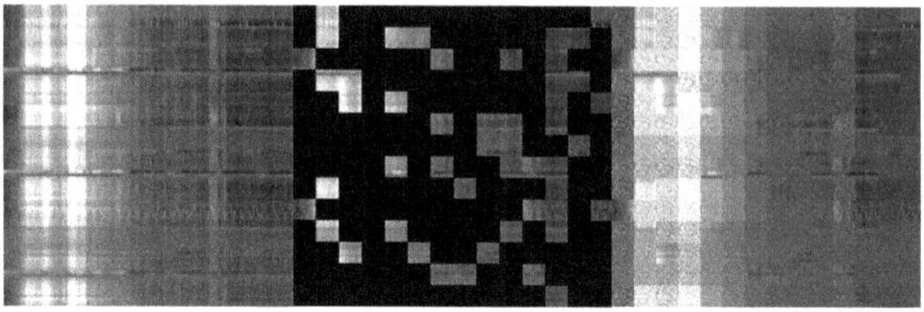

(a) (b) (c)

Figure 6. BERT masking experiments. (**a**) Original image; (**b**) after 75% masking; (**c**) restored images.

5.2. Pre-Training with Transfer Learning

After expanding the solar radio spectrum image dataset there are still few experimental data, therefore, the effect of training the mask model with these data, as shown in Figure 3, is poor. Therefore, we adopt the transfer learning training method. First, we train the mask model on the large dataset ImageNet [31]. At this stage, we do not need to use the tags in the ImageNet dataset. Then, we remove the encoder module and connect it to the fully connected layer and transfer it to the small dataset of the solar radio spectra for fine-tuning. An accuracy comparison using transfer learning is shown in Table 2.

Table 2. Transfer learning effect.

Method	Accuracy (%)
Training from the beginning	70.60
Transfer learning	98.63

As seen in Table 2, the transfer learning ability of the model is very good, which greatly improves the accuracy of the experiment. This is also because the Vision Transformer (ViT) structure requires a large amount of training data. Although the ViT model achieves good results, it is based on the use of a larger dataset than ImageNet. If only ImageNet is used, the results of the ViT model are not better than those of the CNN structure model. However, because the self-supervised method can learn more information about images than the supervised method, the information contained in the image itself is also far greater than the information contained in the label, therefore, the transfer learning effect of the self-supervised model is very good.

5.3. Data Enhancement in Training

In the process of ViT training, to further alleviate overfitting and accelerate convergence, we used many data enhancement methods. In our experiment, we tried color change, random erasure, mixup, cutmix, and their combination. The experimental results are shown in Table 3.

Table 2 shows that based on the accuracy of 98.63% after transfer learning, mixup and cutmix improve the accuracy, the other two methods have no significant improvement effect on the final accuracy, and changing the color even makes the accuracy decrease slightly. Experiments were also conducted to determine how mixup generates virtual data. The experiments were divided into four groups. The accuracy of the model is higher with three interpolations plus the source of the same batch. Experiments were also conducted for the source of cutmix data, which were divided into two groups. The experimental results show that the results are better when the data come from the same batch. The final combination of mixup and cutmix yielded an accuracy of 99.3%, as shown in Table 3.

Table 3. Results of data enhancement experiments.

	Mixing Method		Accuracy (%)	Recall (%)
	Random Erasing		96.7	97.8
Mixup	Color change		98.6	98.4
	+Quadratic interpolation	+Random	98.7	97.3
	+Triple interpolation	+Batch	98.1	97.8
	+Quadratic interpolation	+Batch	98.7	98.4
	+Triple interpolation	+Random	98.9	97.6
Cutmix	+Batch		98.8	97.0
	+Random		98.7	97.3
Mixup + Cutmix			99.3	98.4

5.4. *Dropout*

We conducted experiments on different dropout methods, including dropout, DropPath, and DropAttention [32]. Dropout discards nodes in the network with a certain probability, DropPath discards paths in the network, and DropAttention discards the attention weight in the Transformer with a certain probability. The three methods can prevent overfitting, overcome network degradation and improve the network effect. The best results are shown in Table 4. In the three dropout experiments, DropPath is relatively good and the drop probability is 0.1.

Table 4. Results of dropout experiments.

Method	Accuracy (%)	Recall (%)
Dropout	98.8	97.8
DropPath	98.9	98.4
DropAttention	98.6	98.4

5.5. *Final Classification Results*

After integrating the above methods, the final results of the model and the change curve of its loss are shown in Figure 7. The final accuracy of the model is 99.5%. Compared with the results of some previous studies, it showed a good improvement effect and good migration results are achieved. Although the final restored image is relatively fuzzy, the restored image exhibits the characteristics of the solar radio burst images well. This shows that the network learns the characteristics of solar radio burst spectrum images well.

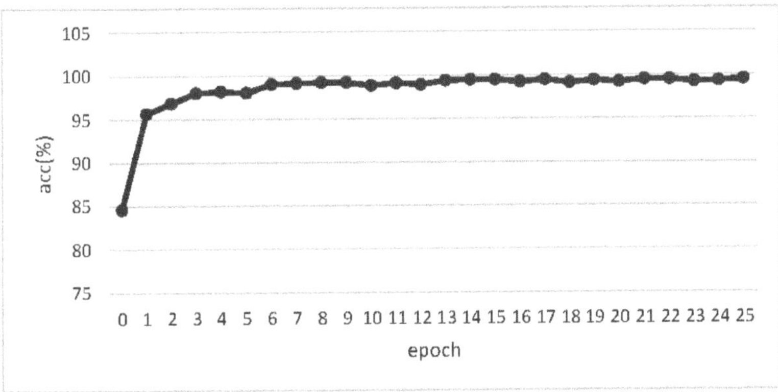

(a)

Figure 7. *Cont.*

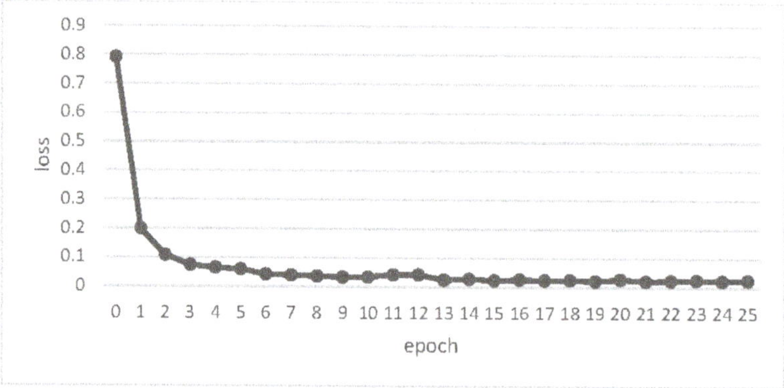

(**b**)

Figure 7. Model training results. (**a**) Accuracy curve, (**b**) Loss curve.

To measure the effectiveness of our model from various aspects, other indicators were considered in the experiment to evaluate our model. First, the confusion matrices of the three categories are calculated and the results are shown in Figure 8. Then, the confusion matrix is used to calculate the precision, recall and specificity of each category. In addition, the F1 score of each category is calculated. An F1 score of 1 is the best and 0 is the worst. The results are shown in Table 5.

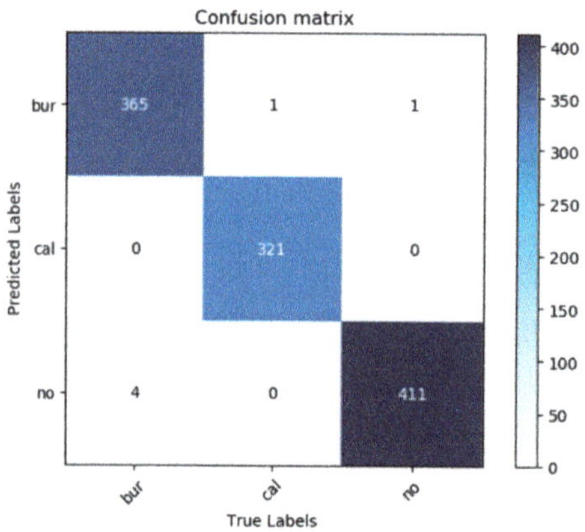

Figure 8. Confusion matrix.

Table 5. Experimental results of our model.

Type	Precision (%)	Recall (%)	Specificity (%)	F1 Score
Burst	99.5	98.9	99.7	0.992
Non-burst	99.0	99.8	99.4	0.994
Calibration	100	99.7	100	0.998

Our model is also compared with other models. The comparison model uses Vision Transformer, Swin Transformer, VGG, GoogLeNet, MoblieNet, ResNet and DenseNet as the core of the experimental network. For a fair comparison, other models are also migrated to the solar radio spectrum image for fine adjustment after pre-training. The results are shown in Table 6.

Table 6. Comparison experiments.

Models	Accuracy (%)	Recall (%)
Vision Transformer	94.3	97.5
VGG16	96.0	98.4
Swin Transformer	99.0	99.1
Resnet	98.9	98.9
MobileNet	95.6	98.1
GoogLeNet	96.6	98.4
DenseNet	99.1	99.1
Ours	99.5	99.7

The self-supervision model can achieve the same accuracy as the current mainstream CNN and Transformer models. At the same time, compared with Vision Transformer, the accuracy of our model is obviously higher under the same transfer learning conditions. This shows that self-supervised learning is indeed more conducive to transfer learning than supervised learning.

For this application, we pay more attention to solar radio bursts that occurred with low probability. It is more critical to find all solar radio bursts as far as possible. Therefore, a high recall rate is important. Our method has achieved a 99.7% recall rate, which is better than other models.

6. Conclusions

In this paper, we propose a solar radio spectrum classification method based on self-supervised learning. By referring to the BERT method in natural language processing, it is improved for solar radio spectrum classification. After randomly masking the solar radio spectrum image, the method lets the model learn to restore the image to learn the image features. This method uses the image itself as a label to enable the network to learn more information, therefore, it is very suitable for transfer learning, thus addressing the issue of fewer solar radio spectrum image datasets. This method can also obtain a good result on a small dataset. Through experiments, an accuracy of 99.5% was achieved on the solar radio spectrum dataset. Compared with other models, our model achieves better experimental accuracy. However, our model has a larger number of parameters and requires more training time. Therefore, we need to continue to study how to reduce the scale of the model.

According to the spectral morphology of solar radio bursts, they can be divided into type I, II, III, IV, V and their associated fine structures. Different solar radio bursts correspond to different solar physical phenomena. Next, we will further subdivide and label the burst samples and use the fine-grained method to classify type I, II, III, IV, V bursts.

Author Contributions: S.L. proposed the network architecture design and the framework; J.C. and S.L. collected and preprocessed the datasets; S.L. performed the experiments; J.C. and S.L. analyzed and discussed the experimental data; S.L. and G.Y. wrote the article; H.Z. and C.T. revised the article and provided valuable advice for the experiments. All authors have read and agreed to the published version of the manuscript.

Funding: This research was funded by the Natural Science Foundation of China (Grant No. 12263008, 11941003), the MOST (Grant No. 2021YFA1600500), the Application and Foundation Project of Yunnan Province (Grant No. 202001BB050032), the Yunnan Provincial Department of Science and Technology-Yunnan University Joint Special Project for Double-Class Construction (Grant No. 202201BF070001-005), the Key R&D Projects in Yunnan Province (Grant No. 202202AD080004), the Open Project of CAS Key Laboratory of Solar Activity, the National Astronomical Observatories (Grant No. KLSA202115) and the Postgraduate Innovation Project of Yunnan University (Grant No. 2021Y269).

Data Availability Statement: The data are available at GitHub: https://github.com/filterbank/spectrumcls (accessed on 1 December 2022).

Acknowledgments: We would like to thank the anonymous reviewers and the editor-in-chief for their comments to improve the article. Thanks also to the data sharer. We thank all the people involved in the study.

Conflicts of Interest: The authors declare no conflict of interest.

References

1. Singh, D.; Raja, K.S.; Subramanian, P.; Ramesh, R.; Monstein, C. Automated Detection of Solar Radio Bursts Using a Statistical Method. *Sol. Phys.* **2019**, *294*, 1500. [CrossRef]
2. Du, Q.-F. A solar radio dynamic spectrograph with flexible temporal-spectral resolution. *Res. Astron. Astrophys.* **2017**, *17*, 98. [CrossRef]
3. Liu, Y.; Ren, D.; Yan, F.; Wu, Z.; Dong, Z.; Chen, Y. Performance comparison of power divider and fiber splitter in the fiber-based frequency transmission system of solar radio observation. *IEEE Access* **2021**, *9*, 24925–24932. [CrossRef]
4. Fu, Q. A new solar broadband radio spectrometer (SBRS) in China. *Sol. Phys.* **2004**, *222*, 167–173. [CrossRef]
5. Ma, L.; Chen, Z.; Xu, L.; Yan, Y. Multimodal deep learning for solar radio burst classification. *Pattern Recognit.* **2017**, *61*, 573–582. [CrossRef]
6. Devlin, J.; Chang, M.W.; Lee, K.; Toutanova, K. *BERT: Pre-Training of Deep Bidirectional Transformers for Language Understanding*; NAACL HLT: Minneapolis, MN, USA, 2019.
7. Liu, H.; Yuan, G.; Yang, L.; Liu, K.; Zhou, H. An Appearance Defect Detection Method for Cigarettes Based on C-CenterNet. *Electronics* **2022**, *11*, 2182. [CrossRef]
8. Yang, L.; Yuan, G.; Zhou, H.; Liu, H.; Chen, J.; Wu, H. RS-YOLOX: A High-Precision Detector for Object Detection in Satellite Remote Sensing Images. *Appl. Sci.* **2022**, *12*, 8707. [CrossRef]
9. Yan, F.-B. A broadband digital receiving system with large dynamic range for solar radio observation. *Res. Astron. Astrophys.* **2020**, *20*, 156. [CrossRef]
10. Zhang, P.J.; Wang, C.B.; Ye, L. A type III radio burst automatic analysis system and statistic results for a half solar cycle with Nancay Decameter Array data. *Astron. Astrophys.* **2018**, *618*, 33260. [CrossRef]
11. Krizhevsky, A.; Sutskever, I.; Hinton, G.E. ImageNet classification with deep convolutional neural networks. *Commun. ACM.* **2017**, *60*, 84–90. [CrossRef]
12. Szegedy, C. Going deeper with convolutions. In Proceedings of the 2015 IEEE Conference on Computer Vision and Pattern Recognition (CVPR), Boston, MA, USA, 7–12 June 2015.
13. Graves, A. Long short-term memory. In *Supervised Sequence Labelling with Recurrent Neural Networks*; Springer: Berlin/Heidelberg, Germany, 2012; pp. 37–45.
14. Bengio, Y. Learning deep architectures for AI. *Found. Trends® Mach. Learn.* **2009**, *2*, 1–127. [CrossRef]
15. Chen, S.S. Research on Classification Algorithm of Solar Radio Spectrum Based on Convolutional Neural Network. Master's Thesis, Shenzhen University, Shenzhen, China, 2018.
16. Jalali, S.M.J.; Ahmadian, S.; Kavousi-Fard, A.; Khosravi, A.; Nahavandi, S. Automated deep CNN-LSTM architecture design for solar irradiance forecasting. *IEEE Trans. Syst. Man.* **2021**, *52*, 54–65. [CrossRef]
17. Yan, B.; Wang, X.; Hu, C.; Wu, D.; Xia, J. Asymmetrical and symmetrical naphthalene monoimide fused perylene diimide acceptors for organic solar cells. *Tetrahedron* **2022**, *116*, 132818. [CrossRef]
18. He, K.; Zhang, X.; Ren, S.; Sun, J. Deep residual learning for image recognition. In Proceedings of the 2016 IEEE Conference on Computer Vision and Pattern Recognition (CVPR), Las Vegas, NV, USA, 27–30 June 2016.
19. Salmane, H.; Weber, R.; Abed-Meraim, K.; Klein, K.-L.; Bonnin, X. A method for the automated detection of solar radio bursts in dynamic spectra. *J. Space Weather. Space Clim.* **2018**, *8*, A43. [CrossRef]
20. Vaswani, A. *Attention is All You Need*; NIPS: Long Beach, LA, USA, 2017.
21. Han, K. A survey on visual transformer. *arXiv* **2020**, arXiv:2012.12556.
22. Bhunia, A.K.; Chowdhury, P.N.; Yang, Y.; Hospedales, T.M.; Xiang, T.; Song, Y.-Z. Vectorization and rasterization: Self-supervised learning for sketch and handwriting. In Proceedings of the 2021 IEEE/CVF Conference on Computer Vision and Pattern Recognition (CVPR), Nashville, TN, USA, 20–25 June 2021.
23. Yue, X. Prototypical cross-domain self-supervised learning for few-shot unsupervised domain adaptation. In Proceedings of the 2021 IEEE/CVF Conference on Computer Vision and Pattern Recognition (CVPR), Nashville, TN, USA, 20–25 June 2021.
24. Yang, J.; Alvarez, J.M.; Liu, M. Self-supervised learning of depth inference for multi-view stereo. In Proceedings of the 2021 IEEE/CVF Conference on Computer Vision and Pattern Recognition (CVPR), Nashville, TN, USA, 20–25 June 2021.
25. Reed, C.J.; Metzger, S.; Darrell, T.S.; Keutzer, K. Selfaugment: Automatic augmentation policies for self-supervised learning. In Proceedings of the 2021 IEEE/CVF Conference on Computer Vision and Pattern Recognition (CVPR), Nashville, TN, USA, 20–25 June 2021.
26. Dosovitskiy, A. An image is worth 16×16 words: Transformers for image recognition at scale. *arXiv* **2020**, arXiv:2010.11929.
27. Liu, Z. Swin Transformer: Hierarchical Vision Transformer using Shifted Windows. In Proceedings of the 2021 IEEE/CVF International Conference on Computer Vision Workshops (ICCVW), Montreal, BC, Canada, 11–17 October 2021.

28. Yuan, L. Tokens-to-Token ViT: Training Vision Transformers from Scratch on ImageNet. In Proceedings of the 2021 IEEE/CVF International Conference on Computer Vision Workshops (ICCVW), Montreal, BC, Canada, 11–17 October 2021.
29. Wang, W. Pyramid Vision Transformer: A Versatile Backbone for Dense Prediction without Convolutions. In Proceedings of the 2021 IEEE/CVF International Conference on Computer Vision Workshops (ICCVW), Montreal, BC, Canada, 11–17 October 2021.
30. Zhu, X.; Su, W.; Lu, L.; Li, B.; Wang, X.; Dai, J. Deformable DETR: Deformable Transformers for End-to-End Object Detection. *arXiv* **2020**, arXiv:2010.04159.
31. Deng, J.; Dong, W.; Socher, R. Imagenet: A large-scale hierarchical image database. In Proceedings of the 2009 IEEE Conference on Computer Vision and Pattern Recognition, Miami, FL, USA, 20–25 June 2009.
32. Zehui, L.; Liu, P.; Huang, L. DropAttention: A Regularization Method for Fully-Connected Self-Attention Networks. *arXiv* **2019**, arXiv:1907.11065.

universe

MDPI

Article

The Analysis and Verification of IMT-2000 Base Station Interference Characteristics in the FAST Radio Quiet Zone

Jian Wang [1,2,3], Yibo Zhao [1], Cheng Yang [1,2,*], Yafei Shi [1,2,*], Yulong Hao [1], Haiyan Zhang [4,5,6], Jianmin Sun [7] and Dingling Luo [8]

1 School of Microelectronics, Tianjin University, Tianjin 300072, China; wangjian16@tju.edu.cn (J.W.); haoyulong98@tju.edu.cn (Y.H.)
2 Qingdao Institute for Ocean Technology, Tianjin University, Qingdao 266200, China
3 Shandong Engineering Technology Research Center of Ocean Information Awareness and Transmission, Qingdao 266200, China
4 National Astronomical Observatories of China, Chinese Academy of Sciences, Beijing 100101, China
5 National Key Laboratory of Radio Astronomy and Technology, Beijing 100101, China
6 Hebei Key Laboratory of Radio Astronomy Technology, Shijiazhuang 050081, China
7 Radio Monitoring Station of Guizhou, Guiyang 550001, China
8 Guizhou Province Network and Information Security Evaluation and Certification Center, Guiyang 550001, China; noah_leo@163.com
* Correspondence: ych2041@tju.edu.cn (C.Y.); shiyafei@tju.edu.cn (Y.S.)

Abstract: In this study, we aim to analyze the electromagnetic interference (EMI) regarding the Five-hundred-meter Aperture Spherical radio Telescope (FAST) caused by base stations in the International Mobile Telecommunications-2000 (IMT-2000) frequency band. By analyzing the frequency bands used by the transmitting and receiving devices and the surrounding environmental parameters and utilizing an approach to predicting radio wave propagation loss that is based on deterministic methods, we conclude by comparing the predicted received power at the FAST with its interference protection threshold. Our analysis demonstrates that, currently, only 55.31% of IMT-2000 base stations in the FAST radio quiet zone (RQZ) meet the protection threshold. Additionally, this article verifies the applicability and accuracy of the radio wave propagation model used in the research based on field strength measurements. Overall, this study provides valuable insights for improving the electromagnetic environment surrounding FAST and reducing the EMI caused by mobile communication base stations. It also provides corresponding analysis methods and useful suggestions for analyzing electromagnetic radiation interference in other radio telescopes.

Keywords: radio telescopes; electromagnetic interference; IMT-2000; FAST

1. Introduction

The Five-hundred-meter Aperture Spherical radio Telescope (FAST) was approved in 2007 and completed acceptance testing in 2020, officially entering its operational phase. As a major national advanced scientific and technological project in China, it holds the record for being the largest spherical radio telescope with a single-dish configuration and the most exceptional sensitivity worldwide [1–3]. FAST, known as the "Chinese Eye of the Sky", has a comprehensive performance about 10 times higher than that of the Arecibo telescope in the United States, which was destroyed in December 2020 [4]. Since its completion, FAST has achieved many significant scientific research results [5,6]. As a passive observational device for radio astronomy, radio telescopes rely on high-sensitivity terminal equipment to receive radio signals from celestial bodies for scientific research. Therefore, the radio astronomy service is highly vulnerable to electromagnetic interference (EMI) from other active services, which may affect its inherent scientific observations [7]. In addition, to strictly control the electromagnetic environment around FAST, the local government has established a radio quiet zone (RQZ) centered on the FAST site, with a radius of 30 km.

Among them, 0–5 km is the core area, 5–10 km is the middle area, and 10–30 km is the remote area. However, numerous mobile communication base stations have yet to meet the daily communication needs of local residents. These base stations may interfere with FAST during operation and affect its performance. Therefore, it is crucial to monitor the electromagnetic environment surrounding FAST, and exploring the signal radiation pattern of mobile communication base stations is essential to ensure the inherent observation of FAST.

Due to the ultra-high sensitivity of FAST, it is susceptible to electromagnetic signals. To mitigate interference from other businesses, it is necessary to implement electromagnetic compatibility and interference avoidance measures before construction and operation [8,9]. For example, Wang et al. have established a FAST satellite interference mitigation system for satellite interference sources. This system uses specific antennas to detect satellite interference in the 1–5 GHz frequency band and establishes a satellite EMI database to suppress satellite interference. Through practical observation, the feasibility of this system has been verified, and a practical solution has been provided to mitigate the impact of satellite interference on FAST [10]. Wang et al. analyzed the fading characteristics in the UHF band of the karst landforms around the FAST site based on the experimental datasets of six frequency channels around FAST using the Kolmogorov–Smirnov statistical method [11]. Additionally, based on radio propagation methods and cognitive theory, they also analyzed the radiation characteristics of mobile communication base stations in the 870–878.6 MHz frequency band within the FAST RQZ. Four strategies for interference avoidance and frequency coordination were proposed for base stations exceeding the FAST interference protection threshold, and the consistency of these strategies was verified through experimentation. In addition, a set of evaluation criteria for frequency coordination strategies was established to analyze the satisfaction of the proposed strategy under different criteria. This research provides theoretical and experimental support for improving the radiation interference of mobile communication base stations within the FAST RQZ [12].

In order to analyze the potential EMI caused by International Mobile Telecommunications-2000 (IMT-2000) mobile communication base stations on the operation of FAST, we use the method listed in Recommendation ITU-R. Based on the parameters of equipment and frequency information at the transceiver end, combined with the protection requirements of radio astronomy services, we analyze whether the IMT-2000 base station in the FAST RQZ will generate EMI to FAST.

2. Analysis Object

2.1. FAST

The location of FAST is situated in Kedu Town, Dawodang, within Pingtang County, located in the Guizhou Province of China (106.85° E, 25.64° N). With its natural geographic advantages and the unique structure designed by Chinese scientists, it is currently the world's largest and most sensitive single-dish radio telescope and a major scientific and technological infrastructure project of China's Eleventh Five-Year Plan. The scientific objectives of FAST mainly include: observing neutral hydrogen in the universe, searching for new pulsars, measuring very long baseline interference (VLBI), and extending deep space communication capabilities [13]. The FAST receiving and terminal systems currently include seven sets of receivers involving 70 MHz–3 GHz [14]. As shown in Figure 1, the IMT-2000 base stations around FAST include two operators, such as China Mobile Communications Corporation (CMCC) and China Unicom Communications Corporation (CUCC), mainly distributed within a range of 5–30 km from FAST, as radio equipment is prohibited in the core area of RQZ.

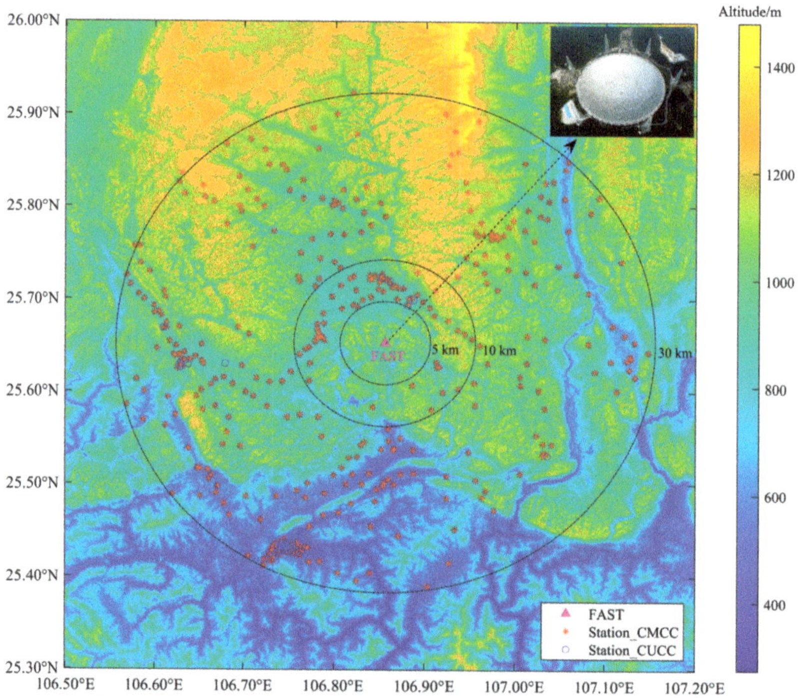

Figure 1. The distribution of IMT-2000 base stations around FAST.

According to the interference protection requirements for FAST and Recommendation ITU-R RA.769 [15], IMT-2000 base stations mainly involve two types of receivers in FAST, namely, B05 and B07. The operating frequency range of the B05 receiver is 1100–1900 MHz, and the system noise temperature is 20 K, while the operating frequency range of the B07 receiver is 2000–3000 MHz, and the system noise temperature is also 20 K. The relevant interference threshold values are shown in Figure 2. Furthermore, in the FAST-IMT interference analysis, we are particularly concerned with the maximum input power that the FAST can tolerate, and the B05 and B07 receiver protection power thresholds are both −199 dBW.

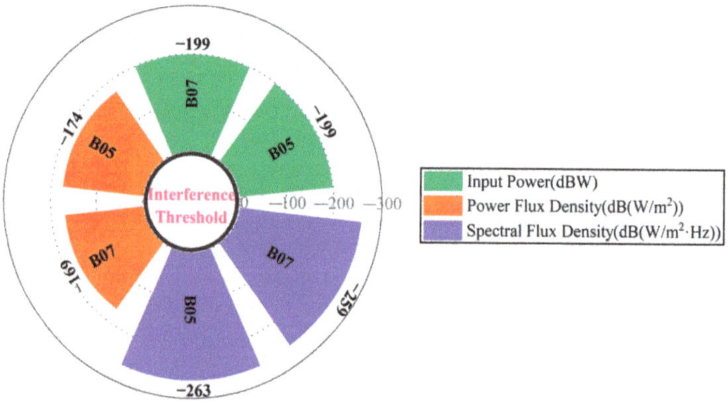

Figure 2. The interference protection requirements for FAST.

In analyzing interference on the FAST-IMT link, the FAST antenna serves as the receiving end, and its antenna gain in the direction of the received signal is the primary consideration. The typical antenna radiation patterns of the FAST antenna include the main vertical plane (zenith angle is 0°), the central horizontal plane (elevation angle is 0°), and the three-dimensional radiation pattern, as shown in Figure 3. The pattern of the antenna reveals a well-shaped main lobe with a narrow beamwidth, with a peak gain of 74.21 dB, accompanied by a number of minor lobes and back lobes of lower gain, with the highest side lobe located at about 120° with respect to the main lobe, as shown in Figure 3a. Figure 3b shows that the horizontal pattern of FAST is directionless over 360° when the elevation angle is 0°, exhibiting a uniform distribution of equivalent radiated power with a gain of 74.21 dB.

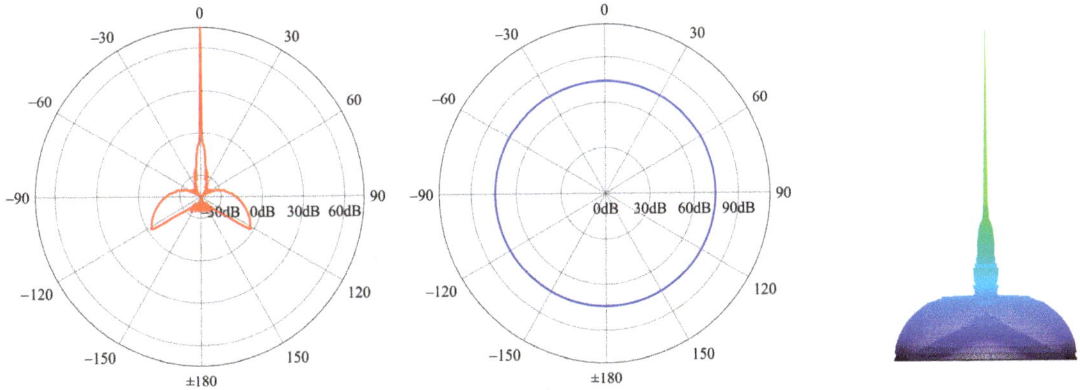

(a) (b) (c)

Figure 3. The typical radiation pattern of the FAST antenna at 1905 MHz: (**a**) 2D-vertical direction (zenith angle is 0°); (**b**) 2D-horizontal direction (elevation angle is 0°); (**c**) 3D.

2.2. IMT-2000 Base Station

The IMT-2000 system was formally proposed by ITU in 2000, which extends the 2G communication system from narrowband to broadband. The IMT-2000 system is mainly committed to mobile internet technology, which fully integrates wireless communication and other communication technologies, such as the Internet. It operates in the 2000 MHz frequency band, with an uplink frequency band of 1890–2030 MHz and a downlink frequency band of 2110–2250 MHz, abbreviated as IMT-2000. The IMT-2000 system mainly adopts Code Division Multiple Access (CDMA) as its core technology and has now formed three mainstream technology standards, including wideband W-CDMA, CDMA-2000, TD-SCDMA, and others. The emergence of the IMT-2000 system has made mobile communication services capable of multimedia transmission, better transmission quality, and higher spectral efficiency [16].

Wideband Code Division Multiple Access, referred to as W-CDMA, is a technology that evolved from the GSM core network. It primarily employs CDMA with a 5 MHz bandwidth, fast power control for both uplink and downlink, downlink transmit diversity, and synchronous and asynchronous operation between base stations. W-CDMA is the most widely adopted, has the most diverse range of terminal equipment among the IMT-2000 standards, and is currently operated mainly by CUCC. CDMA-2000 is a wideband CDMA technology developed from the narrowband CDMA-2000 technology. The forward link can adopt transmit diversity and forward/backward power control to improve channel fading resistance and channel capacity. It is fully compatible with the CDMA-2000 system and inherits the transmission characteristics of the CDMA-2000 system. It is currently operated by CTCC. TD-SCDMA technology was proposed by China to ITU in 1999. It integrates intelligent wireless, synchronous CDMA, software-defined radio, and other related technologies. It has significant advantages in terms of business support, spectrum utilization, switching efficiency, anti-interference, and compatibility. It is highly valued by communication equipment manufacturers and is primarily operated by CMCC, as shown in Figure 4.

Figure 4. The frequency distribution of each operator (IMT-2000 frequency band).

The frequency bands used in the FAST RQZ for IMT-2000 include four frequency bands, namely, 1880–1920 MHz, 2010–2025 MHz, 2110–2130 MHz, and 2130–2155 MHz. In the analysis of interference characteristics of FAST-IMT, the EMI from the base station, which serves as the transmitting end, consists of two parts: electromagnetic leakage interference from the transmitter equipment and radiation interference from the IMT-2000 base station antenna [2]. Compared with the radiation interference from the base station antenna, the leakage interference from the transmitter equipment is weaker. Therefore, when analyzing the EMI of base stations, the radiation interference from the base station antenna is primarily considered. According to the interference prediction method described in Section 3.1, to predict the interference generated by communication base stations, relevant parameters such as the baseband resource reference power of the base station antenna, antenna gain, antenna height, and antenna radiation pattern need to be taken into account. All IMT-2000 base stations in the FAST RQZ use the same antenna model, which operates in 880–960/1700–2200/1880–2700 MHz frequency bands, with an azimuth of 65° and a tilt angle of 0°. Figure 5 shows the typical directional patterns of the IMT-2000 base station antenna in the working frequency bands, including the central horizontal plane (elevation angle is 0°), main vertical plane (zenith angle is 0°), and three-dimensional radiation pattern. The base station antenna exhibits a relatively wide main beam with a low gain but also shows significant sidelobes that may contribute to cross-polarization and intermodulation interference. Additionally, the minimum gain of the antenna is −30 dB, and the peak gain is 0 dB.

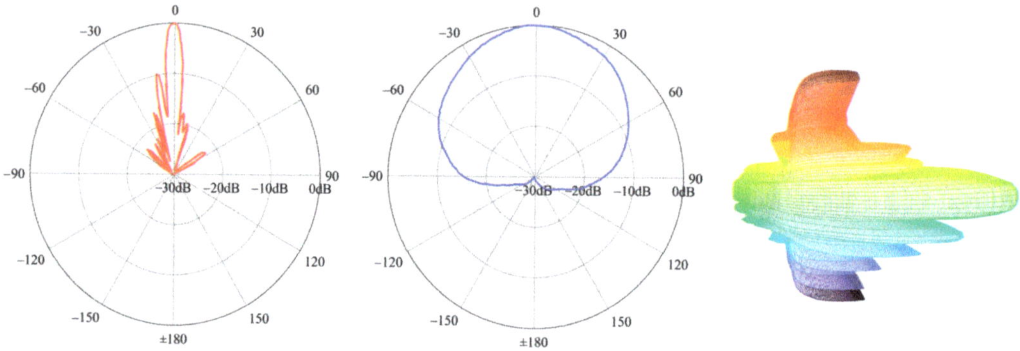

(a) (b) (c)

Figure 5. *Cont.*

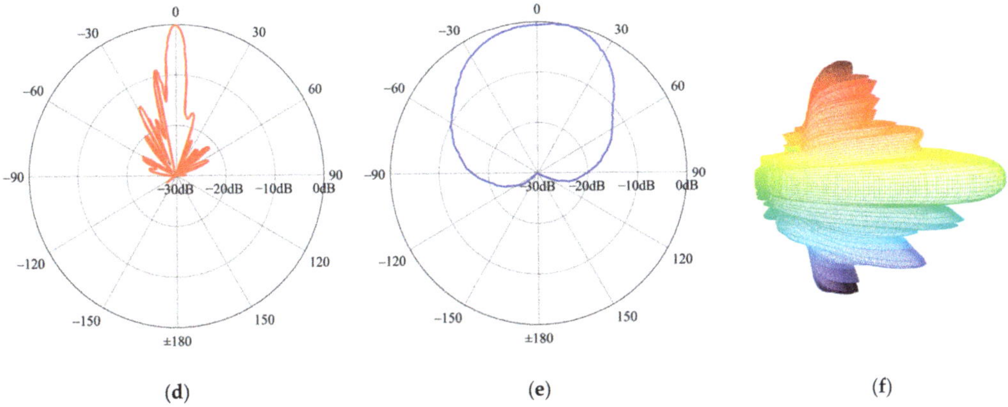

Figure 5. The typical radiation pattern of the base station antenna at 1700–2200 MHz and 1880–2700 MHz: (**a**) 2D-vertical direction (1700–2200 MHz); (**b**) 2D-horizontal direction (1700–2200 MHz); (**c**) 3D (1700–2200 MHz); (**d**) 2D-vertical direction (1880–2700 MHz); (**e**) 2D-horizontal direction (1880–2700 MHz); (**f**) 3D (1880–2700 MHz).

3. EMI Analysis

3.1. Analysis Method

EMI may exist at any time due to one or more propagation mechanisms, and which propagation mechanism is dominant mainly depends on the current weather conditions, the terrain environment, the radio frequency used, the percentage of time of interest, and the propagation path distance, among others. In the problem of interference prediction, it is difficult to propose a unified and consistent method for different distances and time percentages. Therefore, to ensure prediction accuracy, appropriate propagation models should be selected for different climate and path conditions [17].

In radio astronomy, the power I received from the interfering source during observation time T can be expressed as [18]

$$I = \frac{1}{N}\sum_{i=1}^{N} \frac{P_t(i) \cdot G_t(i) \cdot G_r(i)}{L_p(i)} \tag{1}$$

where N is the number of samples in the T; $P_t(i)$ is the transmit power level value (W) of the radio astronomy input antenna at instant i; $G_t(i)$ is the gain of the interference source antenna in the direction of the receiving antenna at instant i; $G_r(i)$ is the gain of the receiving antenna in the direction of the transmitter at instant i; $L_p(i)$ is the transmission loss at instant i. Here, N, $P_t(i)$, $G_t(i)$, and $G_r(i)$ are all set parameters, and $L_p(i)$ can be predicted by the corresponding propagation model.

In the interference prediction analysis of FAST-IMT, ITU-R P.2001 is employed to predict the basic transmission loss caused by signal enhancement and fading. This model can be over distances ranging from 3 km to at least 1000 km beyond the effective range of 0% to 100% annual coverage [19]. The prediction analysis model involves four sub-models, namely, propagation close to the surface of the Earth, anomalous propagation, troposcatter propagation, and ionospheric propagation. Finally, a Monte Carlo simulation-based comprehensive method is utilized to predict the total transmission loss along the path, as shown in Figure 6.

(1) Sub-model 1—Propagation close to the Earth's surface: when the height of the transmitting and receiving antenna is low to the ground, with the maximum radiation direction along the surface, radio waves mainly propagate along the surface of the Earth. At this time, the transmission loss mainly includes free-space basic transmission loss, diffraction loss, clear-air effect, atmospheric attenuation, and other factors.

(2) Sub-model 2—Anomalous propagation: atmospheric propagation mainly refers to the anomalous propagation of the atmospheric layers, namely, the atmospheric waveguide phenomenon.
(3) Sub-model 3—Troposcatter propagation: affected by different solar irradiation intensities, thus forming an uneven propagation medium in the troposphere and causing scattering. The troposcatter transmission loss mainly considers the basic troposcatter transmission loss, rain–snow precipitation attenuation, and atmospheric absorption loss.
(4) Sub-model 4—Ionospheric propagation: for long-haul and low-frequency predictive links, it is essential to consider the ionospheric scattering transmission loss caused by sporadic-E, mainly including the one-hop propagation mode and two-hop propagation mode.
(5) Set relevant parameters based on the device and environment: FAST and IMT-2000 base station locations, frequencies, heights of transceiver antennas, environment type, time probability, location probability, and other basic path information parameters. Then, determine the propagation parameters such as the propagation distance, path midpoint position, sea propagation length, path elevation angle, refractive index, precipitation probability, equivalent Earth radius, effective height, and path roughness parameters.
(6) Free-space basic transmission loss: when the transmitting and receiving antennas are located within the "visible" distance from each other, the radio wave propagates point-to-point along a straight line without reflection or scattering.
(7) When the transmitting and receiving antennas are beyond the line-of-sight range, radio waves propagate mainly through diffraction, including diffraction losses for the Earth's spherical surface, Bullington diffraction losses for the actual profile, and a notional smooth profile.
(8) The primary components of the gaseous attenuation effect are total gaseous attenuation occurring during non-rain periods and gaseous attenuation resulting from water vapor in non-rain and rain situations.
(9) The clear-air zero-fade effect under no atmospheric ducting mainly includes the refractive index change effect, the reflectivity of rain clouds, and the atmospheric thermal noise.
(10) The transmit angle-dependent loss is the unique angular attenuation in the irregular propagation mechanism. When the corrected path angular separation is not greater than 0, its loss is 0; otherwise, it is the product of angular attenuation and the corrected total angular separation.
(11) The time-distance-dependent loss depends on the distance and time percentage of the great circle.
(12) According to the longitude and latitude coordinates of the station site, combined with the climatic zone model specified by ITU, determine the climatic zone and obtain the climatic zone's meteorological and atmospheric structure parameters.
(13) Rain–snow and precipitation attenuation loss include the attenuation caused by rain, snow, and rainfall in the typical path between the transmitting and receiving antennas. It can be calculated using iterative functions based on the positions of the transmitting and receiving terminals, the antenna heights, and the path length between the transmitting and receiving antennas.
(14) The dominant factor contributing to atmospheric absorption loss is the combined effect of total gaseous attenuation during non-rain periods and gaseous attenuation resulting from water vapor in both non-rain and rainy situations.

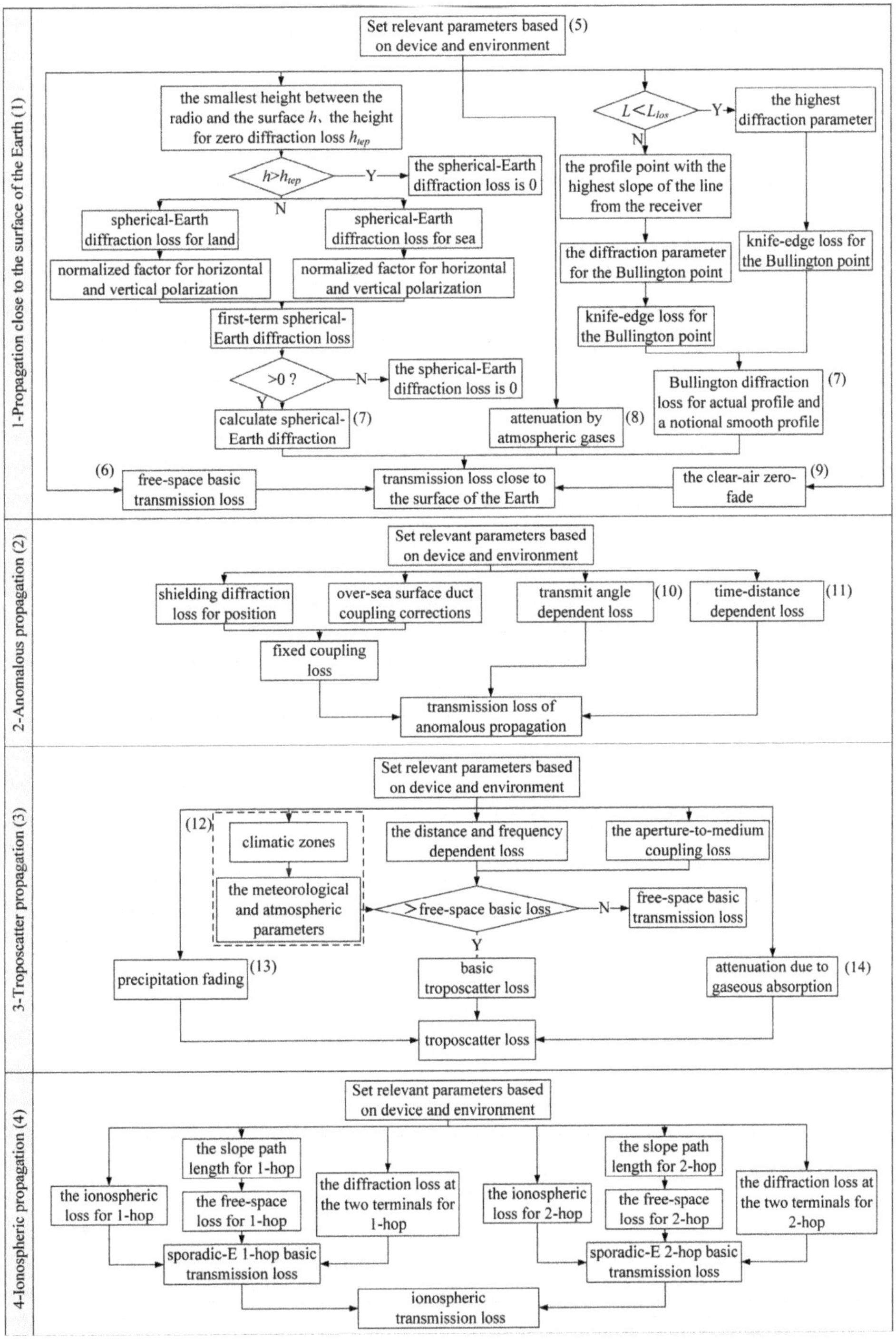

Figure 6. The propagation prediction process of the ITU-R P.2001 model.

3.2. Analysis Result

Using the interference evaluation method described in Section 3.1, and based on the 3 arc-seconds 90 m (SRTM3) high-precision elevation data, the interference caused by the IMT-2000 base stations listed in Figure 1 regarding FAST is analyzed. Because the zenith angle of FAST will change with different observation tasks and the antenna gain towards the direction of receiving signals from the base station will also change, in order to consider all possible scenarios to the maximum extent in the calculation process, several special parameters are set, as follows:

- Frequency: The upper and lower limits of each IMT-2000 base station working frequency band are selected as the analysis main frequency points, including multiple frequencies such as 1880 MHz, 1885 MHz, 1900.4 MHz, 1905 MHz, 1910.4 MHz, 1920 MHz, 2010 MHz, and 2025 MHz under CMCC and 2130.1 MHz, 2131.1 MHz, 2135.1 MHz, 2136.1 MHz, and 2155.1 MHz under CUCC.
- Time percentage: Five kinds of time probability, including 1%, 10%, 50%, 90%, and 99%.
- FAST receiving antenna zenith angle: 0°, 10°, 20°, 30°, and 40°.
- Antenna radiation direction: towards the base station transmitting antenna and away from the base station transmitting antenna.

Because the working frequency of the FAST receiver does not involve 1900–2000 MHz, the communication base station operating in 1900.4 MHz, 1905 MHz, 1910.4 MHz, and 1920 MHz under CMCC will not cause interference regarding FAST. Here, we only analyze the communication base station that shares the working frequency band with the receiver.

Based on the above parameters, we analyzed the interference situation of IMT-2000 mobile base stations, with a total of 62,415 data analyzed, of which 34,521 data meet the requirements, accounting for 55.31%. The evaluation results are classified and counted according to nine distance ranges of 5–6 km, 6–7 km, 7-8 km, 8-9 km, 9-10 km, 10-15 km, 15–20 km, 20–25 km, and over 25 km, as shown in Figure 7 (Note: there are no IMT-2000 base stations of CTCC near the FAST). The results show that: under the current conditions, the proportion of data meeting the FAST threshold requirements for base stations within a distance range of over 25 km is the highest, accounting for 64.03%, while the proportion for the 7–8 km range is the lowest, accounting for only 24.21%. Meanwhile, Table 1 provides the statistical proportion of data that meets the requirements for different receivers of FAST involved in the IMT-2000 frequency band. In the IMT-2000 frequency band, the operating frequency of the communication base station of CUCC is above 2100 MHz, which does not involve the B05 (1100–1900 MHz) receiver. Within the frequency band of the B05 receiver, the analysis data of IMT-2000 base stations meet the FAST protection requirements, accounting for 55.06%. The percentage is 56.00% in the B07 receiver frequency band.

Table 1. The statistical analysis of the proportion of IMT-2000 receivers meeting the FAST interference threshold.

Receiver	CMCC	CUCC	Total Data	Proportion
B05 (1100–1900 MHz)	25,298	-	45,945	55.06%
B07 (2000–3000 MHz)	400	8823	16,470	56.00%
Total	25,698	8823	62,415	55.31%

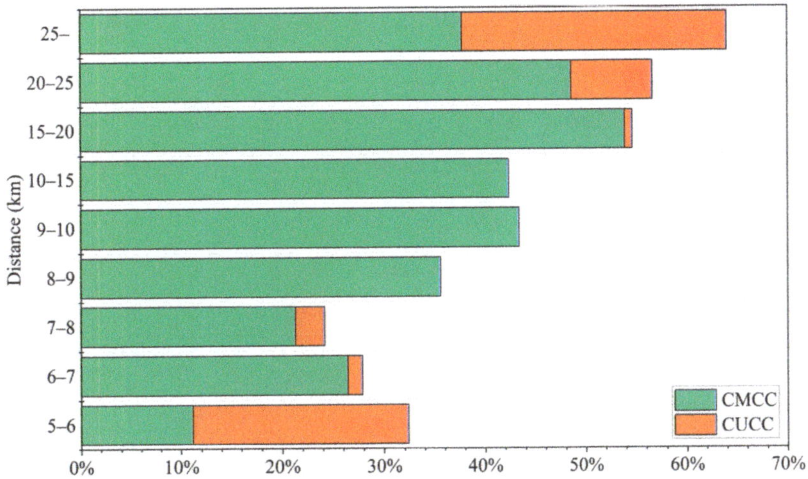

Figure 7. The statistical graph of the proportion of IMT-2000 base stations meeting the FAST interference threshold at various distances.

To suppress the EMI of communication base stations on FAST when the received power exceeds the interference threshold, protective measures can be implemented from the interference source. The most direct and effective measure is to shut down the communication base station, but this method results in the loss of communication quality and has a high cost [20]. In addition, Wang et al. provide four effective measures for mitigating the interference of communication base stations. These measures include increasing the operating frequency of the base station, adjusting the radiation direction of the base station antenna, reducing the height of the base station antenna installation, and decreasing the radiation power of the base station [12]. By implementing these measures, the interference received power of the base station at FAST can be significantly reduced while ensuring communication efficiency and quality.

3.3. Result Validation

In order to verify the applicability and accuracy of the deterministic radio wave propagation model used in this article, based on the frequency parameters of the IMT-2000 base station, the radio propagation characteristics of the propagation link formed between the experimental test base station and the surrounding receiving test positions are predicted. Testing was conducted using calibrated testing equipment and was in accordance with standards such as GB/T 12572-2008, HJ/T 10.2-1996, and GJB2080, and the received power at the receiving point is measured and compared with the simulation prediction results of the propagation model.

Four IMT-2000 base stations, including test base stations 1, 2, 3, and 4, were selected at different directions and distances around FAST. These base stations were operated by different operators and served as test sites. Then, multiple receiving points were set up in different sectors. In order to ensure the accuracy of the test results and avoid interference from terrain, buildings, high-voltage power lines, and other factors, the receiving antenna should be located in flat and open areas. The relationship between the test station and the various receiving test points, as well as the installation of the transmitting and receiving equipment, are shown in Figure 8; the notation "SEC120_1" in the figure refers to the first selected receiving point in sector 120 of the base station, and so on. The specific parameters of the field test are shown in Table 2.

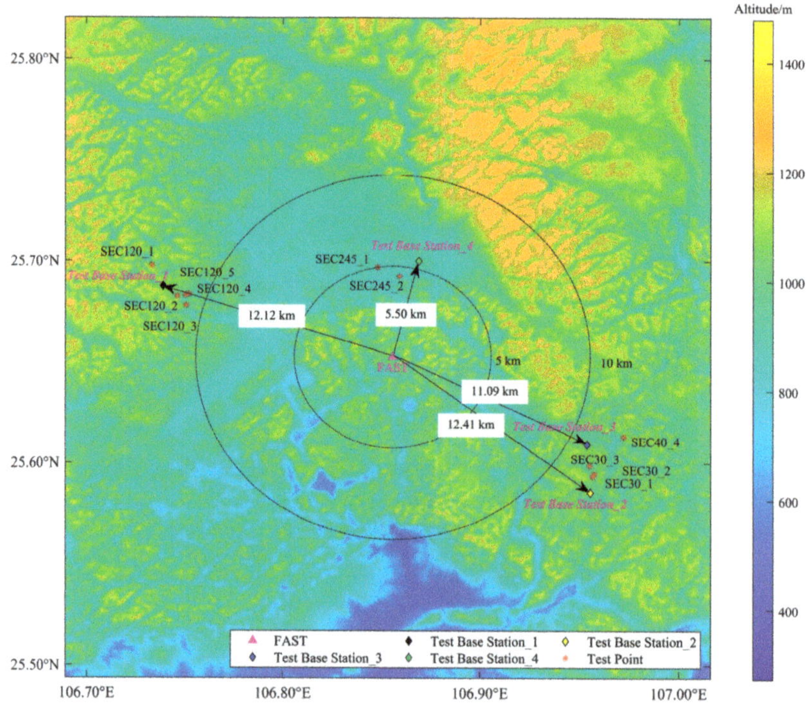

Figure 8. The location information for IMT-2000 test stations.

Table 2. The field test parameter for the IMT-2000 frequency band.

Test Station Parameters							Receiver Point Parameters	
Test Station	**Location**	**Altitude (m)**	**Antenna Height (m)**	**Distance from FAST (km)**	**Frequency (MHz)**	**Transmit Power (W)**	**Antenna Height (m)**	**Path Distance (km)**
1	(106.74° E, 25.69° N)	1150	15	12.12	1895/1909.4			
2	(106.96° E, 25.59° N)	1035	12	12.41	1895	20	4	0.9–2.1
3	(106.95° E, 25.61° N)	960	21	11.09	1895			
4	(106.87° E, 25.70° N)	930	15	5.50	1895			

Taking test base stations 1, 2, 3, and 4 as the main test base stations, we carried out the radio wave propagation characteristics prediction analysis and testing for the IMT-2000 frequency band mobile communication signals of different operators. The statistical results of the predicted link radio wave propagation characteristics between each test station and the surrounding test points are shown in Figure 9 and Table 3. After comparative analysis, the results show that: the predicted results are consistent with the measurement results, with a consistent overall trend. Table 3 shows the mean error (ME), mean deviation (MD), and standard deviation (SD) of four different tests conducted on the base station. Based on the results, test base station 1 had the largest ME of −4.23 dB, indicating a significant underestimation of the received signal power. Test base station 3 had the largest MD and SD values of 10.00 dB, which suggests that the errors for this base station were consistently large and highly variable. The last row presents the total measurement and prediction errors, with an ME of −1.20 dB, an MD of 4.43 dB, and an SD of 5.83 dB. Overall, the total

ME value for all tests is negative, while the MD and SD values are positive, indicating an overall underestimation of the base station performance with relatively good accuracy and variability. However, due to the proximity of several test points near buildings and high-voltage lines near the test base station 3, only the SEC40_4 receiving point was selected. As a result of the small number of test samples, a significant error was observed for the test base station 3.

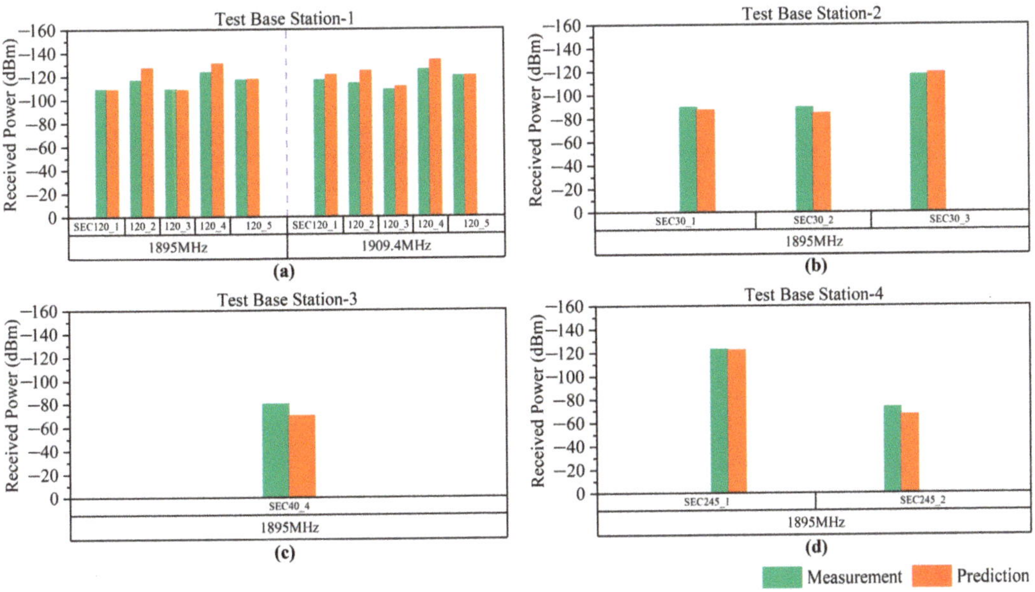

Figure 9. The comparative analysis and statistics between the measurement and prediction: (**a**) Test base station-1; (**b**) Test base station-2; (**c**) Test base station-3; (**d**) Test base station-4.

Table 3. The error analysis of the measurement and prediction.

Test Base Station	ME (dB)	MD (dB)	SD (dB)
1	−4.23	4.49	6.05
2	1.93	2.87	3.23
3	10.00	10.00	10.00
4	3.65	3.65	4.82
Total	−1.20	4.43	5.83

Therefore, the interference evaluation model used in this article can effectively support the analysis of the radio wave propagation characteristics of the link between the IMT-2000 base stations and FAST in the region.

4. Conclusions

This article analyzes whether the IMT-2000 base station in the FAST RQZ affects the normal observation operation of FAST, based on the frequency parameters of FAST and IMT-2000 mobile communication base stations using the ITU-R P.2001 method verified by experiments. By comparing the radio astronomy protection requirements and the protection threshold of FAST, it is concluded that the IMT-2000 base stations in the FAST RQZ meet the threshold data, accounting for only 55.31%. The proportion of data meeting the FAST threshold requirements for base stations within a distance range of over 25 km is the highest, accounting for 64.03%, while the lowest proportion is only 24.21% for the 7–8 km range. Furthermore, the analysis results validated that the predicted ME is −1.20 dB, the MD is

4.43 dB, and the SD is 5.83 dB. Furthermore, the ITU-R P.2001 selected in this article is an ITU recommendation. The universal method can be analyzed more accurately by combining radio meteorological parameters in different regions. The significance of adopting the universal method in this article extends beyond planning mobile communication services in the FAST RQZ and improving the electromagnetic environment around FAST. It also serves as a valuable supplementary approach for other frequency-dependent devices. In the future, the combination of artificial intelligence and other methodologies will be employed to develop localized statistical analysis methods, aiming to provide more robust support for evaluating relevant business operations.

Author Contributions: Conceptualization, J.W. and C.Y.; methodology, software, and validation, J.W., Y.Z., Y.S. and C.Y.; formal analysis, Y.H.; data curation, Y.S. and C.Y.; writing—original draft preparation, Y.Z.; writing—review and editing, J.W., C.Y., Y.S., H.Z., J.S. and D.L.; supervision and project administration, J.W. All authors have read and agreed to the published version of the manuscript.

Funding: This research received no external funding.

Institutional Review Board Statement: Not applicable.

Informed Consent Statement: Not applicable.

Data Availability Statement: Not applicable.

Acknowledgments: Part of this work was supported by the National Astronomical Observatories, the Chinese Academy of Sciences, and the Radio Regulatory Commission of Guizhou Province, China.

Conflicts of Interest: The authors declare no conflict of interest.

References

1. Dalela, C.; Prasad, M.V.S.N.; Dalela, P.K.; Saraf, R. Analysis of WiMAX radio measurements and comparison with some models over dense urban western india at 2.3 GHz. *IEEE Antennas Wirel. Propag. Lett.* **2011**, *10*, 730–733. [CrossRef]
2. Wei, M.; Liang, G.G.; Shi, D.; Gao, Y.G. Base station electromagnetic simulation using ray-tracing method. In Proceedings of the 2012 6th Asia-Pacific Conference on Environmental Electromagnetics (CEEM), Shanghai, China, 6–9 November 2012; pp. 360–362.
3. Yin, J.N.; Jiang, P.; Yao, R. Pose optimization of the FAST feed support system based on the new feed cabin mechanism. *Sci. China Phys. Mech. Astron.* **2023**, *66*, 182–191. [CrossRef]
4. Nan, R.D.; Li, D.; Jin, C.J.; Wang, Q.M.; Zhu, L.C.; Zhu, W.B.; Zhang, H.Y.; Yue, Y.L.; Qian, L. The Five-hundred-meter Aperture Spherical radio Telescope (FAST) project. *Int. J. Mod. Phys. D* **2011**, *20*, 989–1024. [CrossRef]
5. Lin, L.; Zhang, C.F.; Wang, P.; Gao, H.; Guan, X.; Han, J.L.; Jiang, J.C.; Jiang, P.; Lee, K.J.; Li, D.; et al. No pulsed radio emission during a bursting phase of a galactic magnetar. *Nature* **2020**, *587*, 63–65. [CrossRef] [PubMed]
6. Qian, L.; Yao, R.; Sun, J.H.; Xu, J.L.; Pan, Z.C.; Jiang, P. FAST: Its scientific achievements and prospects. *Innovation* **2020**, *1*, 100053. [CrossRef] [PubMed]
7. Sitompul, P.P.; Manik, T.; Batubara, M.; Suhandi, B. Radio frequency interference measurements for a radio astronomy observatory site in Indonesia. *Aerospace* **2021**, *8*, 51. [CrossRef]
8. Zhang, H.Y.; Yao, R.; Hu, H.; Huang, S.J. RFI mitigation on FAST feed cabin. In Proceedings of the 2018 IEEE Global Conference on Signal and Information Processing (GlobalSIP), Anaheim, CA, USA, 26–28 November 2018; pp. 1091–1094. [CrossRef]
9. Zhang, H.Y.; Wu, M.C.; Yue, Y.L.; Gan, H.Q.; Hu, H.; Huang, S.J.; Sun, J.H.; Peng, B.; Nan, R.D.; FAST Collaboration. RFI measurements and mitigation for FAST. *Res. Astron. Astrophys.* **2020**, *20*, 075. [CrossRef]
10. Wang, Y.; Zhang, H.Y.; Hu, H.; Huang, S.J.; Zhu, W.W.; Zhi, G.P.; Zhang, T.; Fan, Z.C.; Yang, L. Satellite RFI mitigation on FAST. *Res. Astron. Astrophys.* **2021**, *21*, 018. [CrossRef]
11. Wang, J.; Shi, Y.F.; Yang, C.; Ji, S.Y.; Su, H.B. Research on fading characteristics of ultrahigh frequency signals in karst landform around radio quiet zone of FAST. *Radio Sci.* **2020**, *55*, e2019RS007048. [CrossRef]
12. Wang, J.; Zhao, Y.B.; Shi, Y.F.; Yang, C.; Hao, Y.L.; Sun, J.M. The electromagnetic compatibility between FAST and public mobile communication stations and its cognitive using frequency strategy. *Res. Astron. Astrophys.* **2022**, *22*, 125005. [CrossRef]
13. Liu, P.; Wang, P.; Li, D.; Zhang, J.; Zhang, L.; Zhang, C.M.; Zhu, W.W.; Yue, D.L.; Dai, S. FAST 19-beam drift-scan pulsar survey simulation. *Prog. Astron.* **2018**, *36*, 173–188. [CrossRef]
14. Nan, R.D.; Zhang, H.Y.; Zhang, Y.; Yang, L.; Cai, W.J.; Liu, N.; Xie, J.T.; Zhang, S.X. Construction progress of the FAST project. *Chin. Astron. Astrophys.* **2017**, *41*, 293–301. [CrossRef]
15. Internatinal Telecommunication Union. *Rec. ITU-R RA.769-2 ITU—R Protection Criteria Used for Radio Astronomical Measurements*; ITU: Geneva, Switzerland, 2020.

16. Zhang, Y.G.; Zhang, W.F.; Zhang, S.H. Overview of the development of 1~6 G mobile communication systems. *China Comput. Commun.* **2020**, *32*, 157–160.
17. Internatinal Telecommunication Union. *Rec. ITU-R P.452-17 ITU—R Prediction Procedure for the Evaluation of Interference between Stations on the Surface of the Earth at Frequencies above about 0.1 GHz*; ITU: Geneva, Switzerland, 2021.
18. Internatinal Telecommunication Union. *Rec. ITU-R RA.1031-3 ITU—R Protection of the Radio Astronomy Service in Frequency Bands Shared with Active Services*; ITU: Geneva, Switzerland, 2021.
19. Internatinal Telecommunication Union. *Rec. ITU-R P.2001-4 ITU—R A General Purpose Wide-Range Terrestrial Propagation Model in the Frequency Range 30 MHz to 50 GHz*; ITU: Geneva, Switzerland, 2021.
20. Huang, S.J.; Zhang, H.Y.; Hu, H. Study on electromagnetic compatibility between FAST and mobile base stations. *J. Deep. Space Explor.* **2020**, *7*, 144–151. [CrossRef]

Disclaimer/Publisher's Note: The statements, opinions and data contained in all publications are solely those of the individual author(s) and contributor(s) and not of MDPI and/or the editor(s). MDPI and/or the editor(s) disclaim responsibility for any injury to people or property resulting from any ideas, methods, instructions or products referred to in the content.

universe
MDPI

Article

A Software for RFI Analysis of Radio Environment around Radio Telescope

Yu Wang [1], Haiyan Zhang [2,3,4,*], Jian Wang [5,6,*], Shijie Huang [2,3], Hao Hu [2,3] and Cheng Yang [5,6]

1 Research Center for Intelligent Computing Platforms, Zhejiang Laboratory, Hangzhou 311100, China
2 National Astronomical Observatories of China, Chinese Academy of Sciences, Beijing 100101, China
3 National Key Laboratory of Radio Astronomy and Technology, Beijing 100101, China
4 Hebei Key Laboratory of Radio Astronomy Technology, Hebei 050081, China
5 School of Microelectronics, Tianjin University, Tianjin 300072, China
6 Qingdao Institute for Ocean Technology, Tianjin University, Qingdao 266200, China
* Correspondence: hyzhang@bao.ac.cn (H.Z.); wangjian16@tju.edu.cn (J.W.)

Abstract: Radio astronomy uses radio telescopes to detect very faint emissions from celestial objects. However, human-made radio frequency interference (RFI) is currently a common problem faced by most terrestrial radio telescopes, and it is getting worse with the development of the economy and technology. Therefore, it is essential to monitor and evaluate interference during the planning, construction, and operation stages of the radio telescope and protect the quiet radio environment around the radio astronomical site. In this paper, we present a software for an RFI analysis of the radio environment around the telescope. In this software, information has been collected, including the location of the site; the technical specifications, such as aperture and the frequency range of the radio telescopes; and the terrain around the site. The software and its modules are composed of telescope, geographic, and meteorological databases, and analysis modules of terrestrial and space-based RFI. Combined with the propagation characteristics of radio waves, we can analyze and evaluate RFI on the ground and in space around the radio telescope. The feasibility of the software has been proved by the experimental implementation of the propagation properties and RFI source estimation. With this software, efficient technical support can be expected for protecting the radio environment around the telescope, as well as improving site selection for planned radio astronomical facilities.

Keywords: radio telescopes; radio frequency interference; RFI analysis software

1. Introduction

Radio astronomers study the Universe by detecting and analyzing radio waves emitted by celestial objects, such as stars, galaxies, and black holes. These signals are transmitted over long distances and require sensitive instruments to detect them. However, RFI from human-made sources, such as mobile phones, TV broadcasts, satellites, and other communication systems, can cause unwanted noise and signal interference, making it difficult to distinguish between the cosmic signals and the RFI. This can seriously degrade the quality of astronomical data, limit the sensitivity of observations, and even render some observations useless.

Therefore, radio astronomers take special care to position their telescopes in remote locations away from human-made RFI sources and use various RFI mitigation techniques to filter out or correct for interference. Setting up the Radio Quiet Zone (RQZ) is the most effective way to protect the electromagnetic environment from the terrestrial RFI [1]. Meanwhile, some large radio telescope facilities have established monitoring systems for terrestrial RFI and developed the satellite RFI databases and prediction systems to prevent interference from satellites [2–4].

In addition, astronomers also need to use some statistical algorithms or manual editing to further mitigate and flag the RFI-contaminated data [5–7]. With new technologies such as multi-beam and phased array feed receivers, the amount of astronomical data is increasing dramatically, and avoiding RFI or providing efficient flagging will be necessary for next-generation radio facilities. Some machine-learning-based RFI recognition methods have been applied and developed to reduce human intervention and increase accuracy [8–10]. This type of supervised learning method requires a large amount of training data to obtain accurate results and is mostly in the experimental stage.

In the field of RFI monitoring and estimation, Ref. [11] provided a scientific basis for the scientific site selection and radio astronomy protection operations through relevant electromagnetic environment tests. Ref. [12] validated the applicability of the ITU-R model in the Karst Region of Guizhou to support the analysis and assessment of the RFI around FAST. Ref. [13] focused on an intelligent monitoring and positioning system to reduce radio frequency interference (RFI) based on monitoring, identifying, and positioning RFI sources. Ref. [14] conducted electromagnetic compatibility studies on FAST, evaluated the RFI impact on mobile communication stations by conducting RFI tests, and proposed a permanent communication station to reduce RFI. Ref. [15] analyzed the radiation characteristics of the public communication stations around FAST and proposed interference avoidance and frequency coordination strategies based on cognitive theory.

In recent years, radio astronomical facilities in China have been rapidly developed. The Five-hundred-meter Aperture Spherical radio Telescope (FAST) has commenced astronomical observation since 2020, the 65-m radio telescope of Shanghai Astronomical Observatory (TM65) has obtained several extraordinary outcomes, and the Qitai 110m Radio Telescope (QTT) in Xinjiang is under construction [16–19]. Meanwhile, the development of the economy and electronic infrastructure near the telescope site has made the electromagnetic environment complex. Wang et al. studied the radio signal's fading characteristics in the Karst landscape environment and analyzed the radiation characteristics of the public communication stations around FAST RQZ [15,20]. In order to manage the surrounding electromagnetic environment more efficiently and balance the requirements of science and economics during the site selection and in the construction and operation phases of the radio telescope, we need to analyze and estimate RFI sources and protect the radio environment around the site.

In this paper, we describe an RFI analysis software that can estimate single or multiple RFI sources on the ground or in space around the radio telescope. Section 2 introduces the software and its modules, including the telescope, geographic, and meteorological database, and the analysis modules of terrestrial RFI and RFI in space. Section 3 presents the experimental implementation of the propagation properties and RFI source estimation. Section 4 gives a conclusion.

2. The RFI Analysis Software of Radio Environment around the Radio Telescope

This section introduces the RFI analysis software of the radio environment around the telescope. It is able to calculate the propagation loss based on the location of the RFI source and receiver, and it can further calculate the field strength and power at the receiver by combining the parameters of the RFI source and receiver. With this information, we can better manage the radio equipment or select the frequency range for observation in the site selection and operation phases of the telescope.

The schematic diagram is shown in Figure 1. It contains the database part and the functional part. The observatory database maintains the location of the site and the technical parameters of each telescope. The geographic and meteorological database gives the Digital Elevation Model (DEM) and radio meteorological environmental data. The RFI sources database offers information on the location of stationary RFI sources, the orbits of sources in space, and their operating parameters. The RFI analysis module provides terrestrial and space-based RFI analysis functions and radio wave propagation characteristics analysis for single or multiple RFI sources.

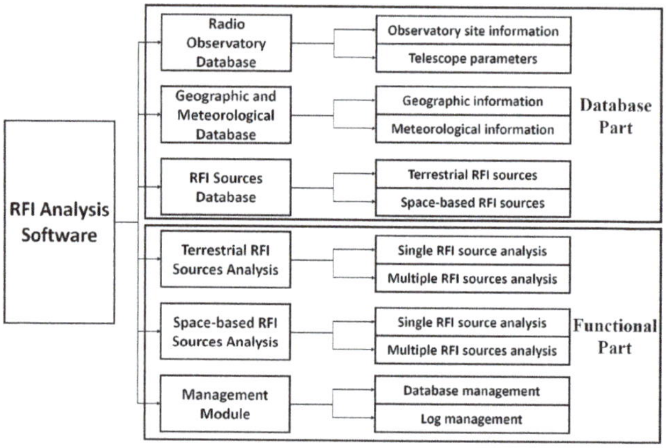

Figure 1. The schematic diagram of the RFI analysis software.

The software is developed on the Visual Studio platform using the C++ development language. It provides a friendly human–machine interface, keeping the interface simple and highlighting the system's main functions. The user interface is divided into two main parts: the menu bar and the map area, where the menu bar provides shortcut buttons for all functions of the software, and the map area carries out station marking and effective display. The software interface is designed in a simple style and can be operated in a guided manner according to the user's functional requirements.

Figure 2 shows the specific architecture of the software, which is designed in a layered approach, with four layers, including the application layer, the core layer, the support layer, and the physical layer.

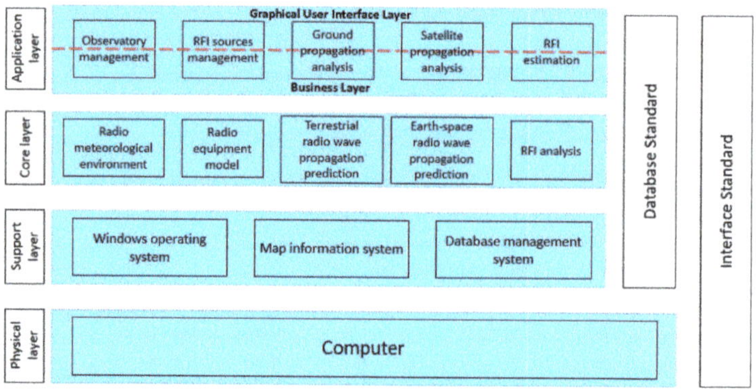

Figure 2. The specific architecture of the RFI analysis software.

The application layer is at the top of the architecture and consists of the graphical user interface layer and the business layer. The graphical user interface layer is responsible for receiving various parameter configuration commands from users and providing a consistent access interface for users, as shown in Figure 3. The business layer provides reusable business function modules, and the implementation of this layer and the graphical user interface layer results in a message-driven mechanism to communicate through a custom set of messages and exchange data information through memory. The functional modules of the application layer mainly include astronomical observatory management, RFI source management, terrestrial point-to-point/area propagation characteristics analysis, satellite point-to-point/area propagation characteristics analysis, etc.

The core layer mainly provides the underlying methods for each functional module of the business layer, including radio meteorological environment modeling, radio equipment modeling, terrestrial radio wave propagation prediction, Earth–Space radio wave propagation prediction, and RFI analysis. This layer provides various kinds of data and analysis results for each functional module. The interaction with the business layer adopts the API interface method, which provides a unified access interface for users and communicates with the upper layer through the interface function.

The support layer provides the required basic service environment for the upper layer, including the Windows operating system, map information system, and database management system. The operating system provides process services, thread services, interface units, etc. The map information system provides the geographic information required by the software. The database management system realizes the storage and management of antenna, equipment data, and radio wave propagation environment data.

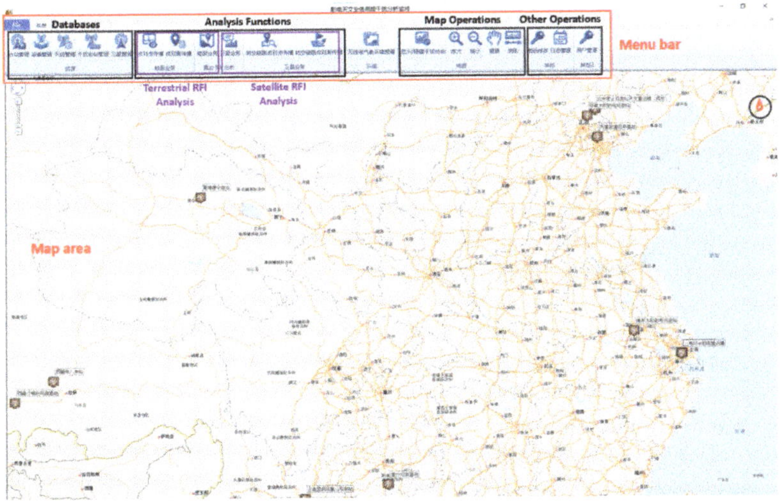

Figure 3. The graphical user interface of the RFI analysis software.

2.1. Databases

The radio observatory database has a hierarchical design, where one observatory may have multiple telescope systems, and one telescope system may also have multiple antennas. It contains the name, the longitude and latitude of the observatory site, and the equipment parameters, including receiving frequency range, aperture, gain, beam pattern, and polarization. Users can add, delete, modify, and query observatory sites, telescope systems, and antenna equipment.

The geographic database provides the 3 arc-sec accuracy DEM with a longitude range from 70° E to 135° E and a latitude range from 10° N to 55° N. A DEM is a 3D computer graphics representation of elevation data to represent terrain or overlaying objects, commonly of a planet, moon, or asteroid. DEMs are used often in geographic information systems and are the most common basis for digitally produced relief maps. The geographic data will be loaded and used to construct the transmission loss model between the RFI transmitter and the receiver.

The meteorological database consists of a ground dielectric constant, conductivity, atmospheric refractive index and gradient, atmospheric temperature, atmospheric humidity, atmospheric pressure, etc. The ground electrical parameters include ground conductivity and dielectric constant, which are determined by measuring the ground wave propagation field strength and time delay from the transmitting source, and then using the ground wave propagation characteristics for inversion. The atmosphere parameters, such as pressure and temperature, are obtained by analyzing meteorological data from 752 terrestrial meteorological stations in China, 120 space-based stations, and more than 600 terrestrial stations in the neighboring areas of China for the past 20 years. The meteorological data required for calculation are mainly selected from the annual statistical data of the ITU, and the parameters are shown in Table 1. Combined with parameters such as the location of the proposed link, antenna height, and percentage time, our software predicts the basic transmission loss not exceeded for a given percentage of an average year based on the radio meteorological data.

Table 1. The parameters in the meteorological database.

No.	Parameter Name	Unit
1	Vertical reflectivity gradient [21]	N-units/km
2	Average sea level value of surface reflectivity [21]	N-units
3	Average annual difference in the values of the reflectivity at the surface and 1000 m above the surface [21]	N-units/km
4	Wet term of the surface reflectivity [21]	ppm
5	Surface water vapor density [22]	g/m^3
6	Mean annual rainfall amount [23]	mm
7	Ratio of convective to total rainfall amount [23]	-
8	Probability of rainy 6-hours periods [23]	%
9	0 degree isotherm height [24]	km
10	Mean rain height [24]	km

The RFI sources database records the terrestrial and space-based RFI sources. Terrestrial RFI data offer the location, frequency range, power, and main lobe angle. Satellite RFI data offer the orbit calculated by the Two-Line Elements (TLE), the antenna beam pattern, the frequency range, and the power. When performing RFI analysis, users can manually add new RFI sources or select RFI sources from the database.

2.2. RFI Analysis Module

We construct the RFI analysis module with these databases to analyze the propagation characteristics and interference situations for different RFI sources. The RFI analysis module supports the analysis of a single RFI source for a single receiver or an area, which can be used to evaluate additional electronics around the telescopes and provide a basis for site selection, respectively. Moreover, the analysis of RFI sources can be terrestrially fixed or satellite services, as shown in Table 2.

Table 2. Scope of application of different types of wave propagation calculation models.

No.	Model Name	Supported Frequency Range	Services
1	Terrestrial radio wave propagation prediction model [25]	30 MHz∼50 GHz	radio, mobile communication, television, radar
2	Earth–Space radio wave propagation prediction model [26–30]	1 GHz∼55 GHz	communication, navigation, radio, television

The terrestrial model facilitates the prediction of propagation characteristics within the troposphere, spanning a frequency range from 30 MHz to 50 GHz. This model considers the main transmission mechanisms within the troposphere, including propagation close to the surface of the Earth, anomalous propagation due to stratified atmosphere, troposcatter propagation, and propagation via sporadic-E reflection. Utilizing numerical analysis techniques for synthesis, the model enables the prediction of transmission loss [25,31]. A practical propagation prediction model will then be derived by further combining domestic experimental data and analysis results.

Compared to free-space propagation, several propagation effects may require consideration when calculating the propagation loss for Earth–space paths: tropospheric effects (including gaseous absorption, and attenuation and depolarization by rain and other hydrometeors), ionospheric effects (such as scintillation and Faraday rotation), and local environmental effects (including attenuation by buildings and vegetation). Moreover, the prediction methods for Earth–space telecommunication systems vary depending on the specific service involved, including broadcasting-satellite systems [27], maritime mobile systems [28], land mobile systems [29], and aeronautical mobile systems [30]. In the case of space-to-Earth paths for broadcasting systems, the propagation attenuation factor $A(f)$ is calculated by the formula, and the unit is dB [26]:

$$A = A_{bs}(f) + A_{sc}(p,f) + A_{gas}(f) + \sqrt{A_{st}^2(p,f) + [A_{rain}(p,f) + A_{cloud}(p,f)]^2} \quad (1)$$

A_{bs} is the antenna attenuation factor, A_{sc} is the ionospheric atmospheric attenuation, A_{gas} is the atmospheric gases attenuation, A_{st} is the troposphere scintillation attenuation, A_{rain} and A_{cloud} are the attenuation factors of rain and cloud, respectively, f is the frequency, and p stands for the time percentage of each parameter.

In practice, the software will combine models based on the service type and frequency range, RFI source location, and electromagnetic wave propagation mode to calculate the propagation loss and field intensity. The next section gives some experimental RFI analysis for terrestrial and satellite RFI sources.

3. Experimental Implementation of the RFI Analysis Software

In this section, we present the experimental implementation of the software for propagation analysis and RFI estimation. Both the terrestrial and satellite RFI parts include the point-to-point and point-to-area functions. The point-to-point analysis is used to guide the installation and application of radio transmitters. The emission power of the equipment is derived from the measurement in the microwave darkroom, then the measurement results are loaded into the software to obtain the interference level in different locations. The point-to-area analysis is used for radio telescope site selection. It is able to calculate the distribution of the interference level of an existing RFI source to an area. Thus, the part with the lowest interference intensity is selected as the alternative site.

3.1. Terrestrial RFI Analysis

The characteristics analysis of the point-to-point radio wave propagation needs to set the transmitting and receiving point information: the user can choose from the database or add new sites, equipment, and antennas. The input information includes the site location, transmitting equipment operating frequency, power, antenna main lobe azimuth and elevation angle, receiving equipment name, sensitivity, and antenna main lobe azimuth and elevation angle. The radio meteorological environment around the transmitting and receiving points is also obtained from the database, and the results are calculated using the algorithm model.

For multiple-RFI-sources analysis, we integrate the received power over a time period of T. The power received from an interferer during observation can be expressed as follows:

$$I = \frac{1}{N}\sum_{i=1}^{N} \frac{P_t(i) \cdot G_t(i) \cdot G_r(i)}{L_p(i)} \quad (2)$$

where I is the interference power in the reference bandwidth at the receiver input averaged over the observation period T, N is the number of samples in the integration time T, $P_t(i)$ is the transmitting power level in the radio astronomy service bandwidth at the input to the antenna, $G_t(i)$ is the gain of the transmitting antenna in the direction of the radio astronomy antenna, $G_r(i)$ is the gain of the radio astronomy antenna in the direction of the transmitter, and $L_p(i)$ is the propagation loss at instant i.

Table 3 and Figure 4 show the computing results of the point-to-point propagation characteristics analysis. The example uses a communication station as the RFI source, which operates at 870 MHz and is located at (106°37′4.0005″ E, 25°37′45.010″ N). The left part shows the plot of the analytical results: the variation in field strength and topography versus distance from the transmitter. The right table gives the corresponding computing results. Compared with the free-space propagation loss, our algorithm further takes into account the terrain and atmosphere, and the calculated field strength is lower and more in line with the actual situation.

Table 3. The point-to-point analytical results for terrestrial RFI source.

Distance (km)	Altitude (m)	Field Strength (dB μV/m)	Free-Space Propagation (dB μV/m)
0	1206	87.78	107.77
3.79	1104.3	33.03	76.21
7.57	1154.2	31.19	70.18
11.36	1091.7	18.79	66.66
15.15	1198.3	25.4	64.16
18.93	1103.4	20.8	62.23
22.72	1017	21.43	60.64
26.51	823.9	13.64	59.3
30.29	837.9	−13.46	58.14
34.08	903	0.05	57.12

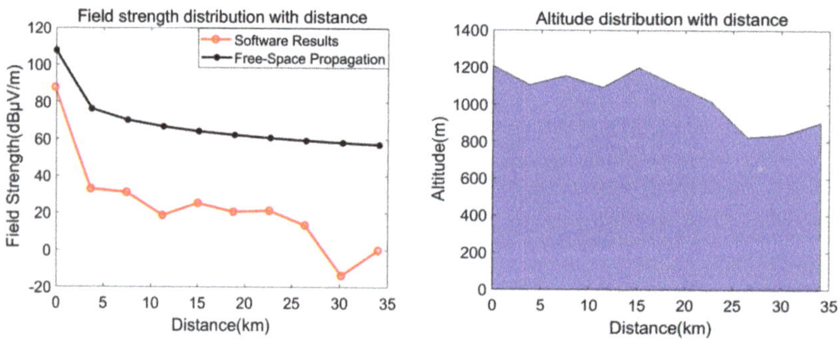

Figure 4. The point-to-point analysis for terrestrial RFI source.

The characteristics analysis of point-to-area radio wave propagation requires setting the transmitting point and receiving area information, where the transmitting point can also be loaded from the database, or added by the user. The receiving area can be obtained by manually entering the latitude and longitude range, or by selecting an area on the map.

Table 4 and Figure 5 show the results of the point-to-area propagation characteristics analysis. The left part gives the distribution of the RFI source field intensity in the selected area. The table on the right gives the corresponding latitude, longitude, altitude, distance from the RFI source, field strength, and propagation loss of the different sampling points in the selected area.

Table 4. The point-to-point analytical results for terrestrial RFI source.

Longitude (°E)	Latitude (°N)	Distance (km)	Altitude (m)	Field Strength (dB μV/m)	Propagation Loss (dB)
106.14119	26.20192	69.885	1744	4.477	169.987
106.14119	26.24359	67.256	1751	5.465	168.97
106.14119	26.28526	64.878	1760	6.457	167.978
106.14119	26.32692	62.739	1771	7.42	167.015
106.14119	26.36859	60.878	1784	8.304	166.13
106.14119	26.41026	59.321	1796	9.129	165.306
106.14119	26.45192	58.092	1804	9.712	164.722
106.14119	26.49359	57.213	1803	9.949	164.486
106.14119	26.53526	56.701	1783	9.718	164.717
...	...	...	...	...	...

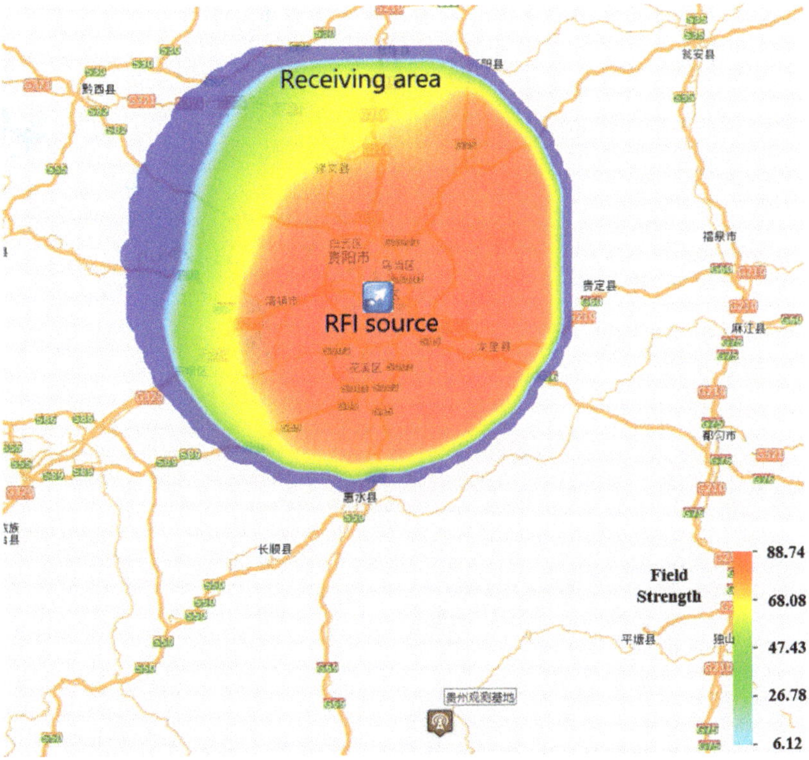

Figure 5. The point-to-area analysis for terrestrial RFI source.

Figure 6 shows the compared results of the software analysis and the practical test. The transmitting site is in Hanglong Town, Guizhou Province, and the receiving site is at the FAST site. The practical test results (black line) cover 0.1–1.7 GHz. It can be seen that there is a high agreement between the analytical results of our model (red line) and the actual measurements, showing that our model is more practical than the free-space propagation model (blue line).

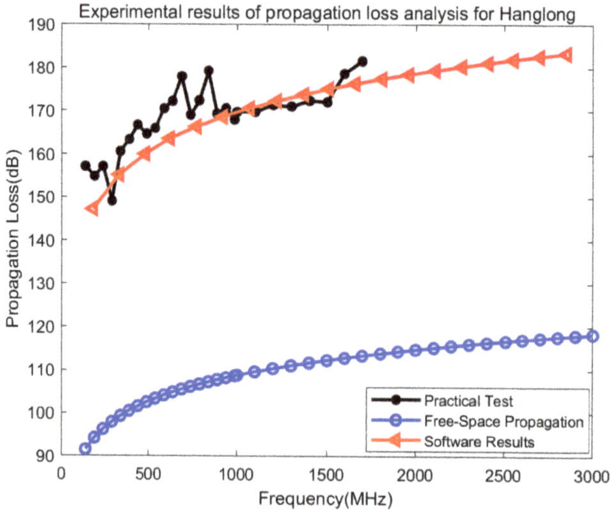

Figure 6. The analytical results of propagation loss from Hanglong to FAST.

3.2. Satellite RFI Analysis

The operation of the satellite RFI analysis function is similar to that of the terrestrial analysis function. Satellite point-to-point radio wave propagation characteristic analysis requires the input of transmitting and receiving equipment information, which can be selected from the database or added manually. The input includes satellite information: name, operating frequency, power, main lobe shape and pointing angle[1,2], and receiver information: name, location, sensitivity, main lobe azimuth and elevation angle. The corresponding radio meteorological parameters are also retrieved from the database, and the results are calculated using the appropriate algorithm model.

Figure 7 shows the analytical results of the satellite RFI point-to-point propagation characteristics. It gives the variations in the transmission losses of GPS_BIIF_4, GPS_BIIF_9, and GPS_BIII_2 to the receiving device with time. Furthermore, the computing results also can be output as a table.

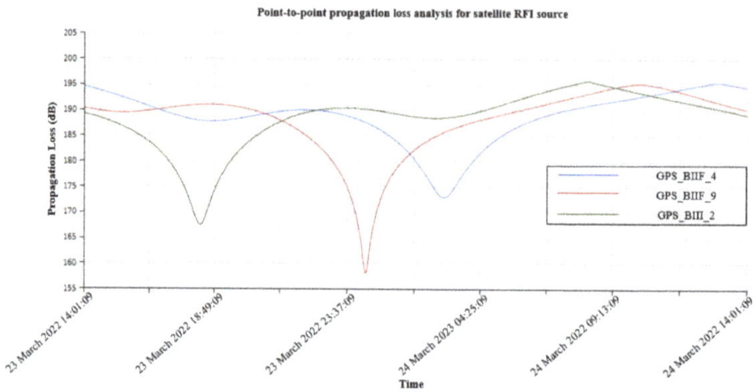

Figure 7. The point-to-point analysis for satellite RFI source.

In reality, interference of the satellite received by radio telescopes is usually the result of multiple satellites acting simultaneously. ITU-R M.1583-1 gives a method for calculating radio telescope interference from non-geostationary orbiting satellites [32]. However, it does not take into account the propagation loss between the satellite and the telescope. Our software combines the estimation method and propagation loss model to give a more realistic estimation.

Figure 8 shows the RFI estimation for multiple satellites including individual satellites and synthetic results. The corresponding time, satellite position, and interference power information can be saved as a table.

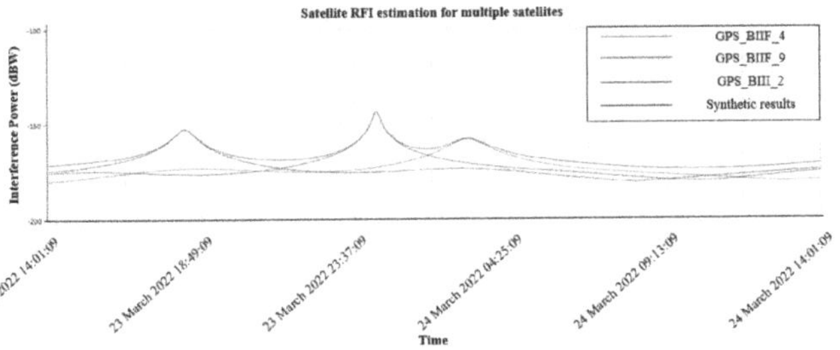

Figure 8. The results of satellite RFI estimation.

In this section, we present the main function of the software, including the propagation loss field strength computing for both terrestrial and satellite RFI sources. Furthermore, we compare the estimated results with actual tests for point-to-point terrestrial RFI sources. The experiments show that our models give identical results to the actual test and are more accurate compared with the free-space model.

4. Conclusions

In conclusion, we have constructed RFI analysis software for the radio environment around the radio telescope, combining complex information from celestial and satellite sources into a database. The database provides RFI sources, radio telescopes, and meteorological and geographic information, and users can add, delete, and modify the database for maintenance operations. The functional module provides the algorithms and interface to calculate the propagation loss for different kinds of RFI sources and environments. We further verified its accuracy by comparing the results of software analysis and practical tests.

With the software, we can use point-to-point analysis to evaluate the new transmitter and guide the subsequent installation and application of the transmitter. On the other hand, the software also provides point-to-area analysis to calculate the interference intensity distribution over an area, which can be useful for the selection of radio astronomical sites. The software is expected to be an efficient tool for protecting the radio environment around the radio telescope. In addition, interference analysis of satellites can guide us in choosing the observation time with minimum impact and support the division of spectrum resources in the future.

Author Contributions: Conceptualization, H.Z. and J.W.; methodology and software, J.W. and C.Y.; validation, Y.W., H.H. and S.H.; writing—original draft preparation, Y.W.; writing—review and editing, H.Z. and J.W.; supervision, H.Z. and J.W.; project administration, H.Z.; funding acquisition, H.Z. All authors have read and agreed to the published version of the manuscript.

Funding: This research was funded by the National Key R&D Program of China, No. 2021YFC2203204, the National Natural Science Foundation of China, No. 12273067 and the National Natural Science Foundation of China: No. 12041301.

Institutional Review Board Statement: Not applicable.

Data Availability Statement: Data sharing not applicable.

Conflicts of Interest: The authors declare no conflict of interest. The funders had no role in the design of the study; in the collection, analyses, or interpretation of data; in the writing of the manuscript; or in the decision to publish the results.

Abbreviations

The following abbreviations are used in this manuscript:

RFI	Radio frequency interference
RQZ	Radio Quiet Zone
API	Application Programming Interface
DEM	Digital Elevation Model
FAST	Five-hundred-meter Aperture Spherical radio Telescope
TLE	Two-Line Elements

Notes

1 https://celestrak.org/, accessed on 22 March 2022.

2 https://www.space-track.org/, accessed on 22 March 2022.

References

1. Baan W.A. Implementing RFI mitigation in radio science. *J. Astron. Instrum.* **2019**, *8*, 1940010. [CrossRef]
2. Zanichelli, A.; Serra, G.; Mack, K.; Nicotra, G.; Bartolini, M.; Cantini, F.; Biaggi, M.D.; Gaudiomonte, F.; Bortolotti, C.; Roma, M.; et al. Towards coordinated site monitoring and common strategies for mitigation of Radio Frequency Interference at the Italian radio telescope. *arXiv* **2022**. arXiv:2207.07571.
3. Katore, S.N.; Raybole, P.A.; Rai, S.; Nayak, S.; Kumar, S. Tools for avoiding satellite interference at the upgraded GMRT. In Proceedings of the 32nd URSI GASS, Montreal, QC, Canada, 19–26 August 2017.
4. Wang, Y.; Zhang, H.Y.; Hu, H.; Huang, S.J.; Zhu, W.W.; Zhi, G.P.; Zhang, T.; Fan, Z.C.; Yang, L. Satellite RFI mitigation on FAST. *Res. Astron. Astrophys.* **2021**, *21*, 18. [CrossRef]
5. Offringa, A.R.; De Bruyn, A.G.; Biehl, M.; Zaroubi, S.; Bernardi, G.; Pandey, V.N. Post-correlation radio frequency interference classification methods. *Mon. Not. R. Astron. Soc.* **2010**, *405*, 155–167. [CrossRef]
6. Sardarabadi, A.M.; van der Veen, A.J.; Boonstra, A.J. Spatial filtering of RF interference in radio astronomy using a reference antenna array. *IEEE Trans. Signal Process.* **2016**, *64*, 432–447. [CrossRef]
7. Zeng, Q.; Chen, X.; Li, X.; Han, J.L.; Wang, C.; Zhou, D.J.; Wang, T. Radio frequency interference mitigation based on the asymmetrically reweighted penalized least squares and SumThreshold method. *Mon. Not. R. Astron. Soc.* **2020**, *500*, 2969–2978. [CrossRef]
8. Akeret, J.; Chang, C.; Lucchi, A.; Refregier, A. Radio frequency interference mitigation using deep convolutional neural networks. *Astron. Comput.* **2017**, *18*, 35–39.
9. Yang, Z.C.; Yu, C.; Xiao, J.; Zhang, B. Deep residual detection of radio frequency interference for FAST. *Mon. Not. R. Astron. Soc.* **2020**, *492*, 1421–1431. [CrossRef]
10. Mesarcik, M.; Boonstra, A.J.; Ranguelova, E.; van Nieuwpoort, R.V. Learning to detect radio frequency interference in radio astronomy without seeing it. *Mon. Not. R. Astron. Soc.* **2022**, *516*, 5367–5378. [CrossRef]
11. Li, J.B.; Peng, B.; Liu, D.L. Site spectrum monitoring of electromagnetic environment for large radio telescopes. *Chin. J. Radio Sci.* **2015**, *30*, 378–382.
12. Yang, C.; Wang, J.; You, X.; Haibin, S.U.; Shengyu, J.I.; Zhao, Z. Applicability of ITU-R P.1546 recommendation in typical terrestrial areas of CHINA. *Chin. J. Radio Sci.* **2019**, *34*, 295–301.
13. Wang, Y.; Zhang, H.; Hu, H.; Huang, S.; Gan, H.; Wu, M.; Zhang, X.; Zhu, W. A Proposed RFI Intelligent Monitoring and Positioning System of FAST. In Proceedings of the 2020 XXXIIIrd General Assembly and Scientific Symposium of the International Union of Radio Science, Rome, Italy, 29 August–5 September 2020; IEEE: Piscataway, NJ, USA, 2020.
14. Zhang, H. et al. [FAST Collaboration]. RFI measurements and mitigation for FAST. *Res. Astron. Astrophys.* **2020**, *20*, 111–114.
15. Wang, J.; Zhao, Y.B.; Shi, Y.F.; Yang, C.; Hao, Y.L.; Sun, J.M. The electromagnetic compatibility between FAST and public mobile communication stations and its cognitive using frequency strategy. *Res. Astron. Astrophys.* **2022**, *22*, 125005. [CrossRef]
16. Nan R.D. Five hundred meter aperture spherical radio telescope (FAST). *Sci. China Ser. G* **2006**, *49*, 129–148. [CrossRef]
17. Chen, R.; Zhang, H.; Jin, C.; Gao, Z.; Zhu, Y.; Zhu, K.; Jiang, P.; Yue, Y.; Lu, J.; Zhang, B.; et al. FAST VLBI: Current status and future plans. *Res. Astron. Astrophys.* **2020**, *20*, 74. [CrossRef]
18. Wang, N. Xinjiang Qitai 110 m radio telescope. *Sci. Sin. Phys. Mech. Astron.* **2014**, *44*, 783. [CrossRef]
19. Liu, X.; Liu, T.; Shen, Z.; Qin, S.; Luo, Q.; Cheng, Y.; Gu, Q.; Zhang, T.; Zhu, F.; Liu, S.-Y.; et al. A Q-band Line Survey toward Orion KL Using the Tianma Radio Telescope. *Astrophys. J. Suppl. Ser.* **2022**, *263*, 13. [CrossRef]
20. Wang, J.; Shi, Y.; Yang, C.; Ji, S.; Su, H. Research on Fading Characteristics of Ultrahigh Frequency Signals in Karst Landform Around Radio Quiet Zone of FAST. *Radio Sci.* **2020**, *55*, 10. [CrossRef]

21. Recommendation ITU-R P.453-14, The Radio Refractive Index: Its Formula and Refractivity Data. 2019. Available online: https://www.itu.int/rec/R-REC-P.453/en (accessed on 21 April 2021).
22. Recommendation ITU-R P.836-6, Water Vapour: Surface Density and Total Columnar Content. 2017. Available online: https://www.itu.int/rec/R-REC-P.836/en (accessed on 21 April 2021).
23. Recommendation ITU-R P.837-7, Characteristics of Precipitation for Propagation Modeling. 2017. Available online: https://www.itu.int/rec/R-REC-P.837/en (accessed on 21 April 2021).
24. Recommendation ITU-R P.839-4, Rain Height Model for Prediction Methods. 2013. Available online: https://www.itu.int/rec/R-REC-P.839/en (accessed on 21 April 2021).
25. Recommendation ITU-R P.2001-4, a General Purpose Wide-Range Terrestrial Propagation Model in the Frequency Range 30 MHz to 50 GHz. 2021. Available online: https://www.itu.int/rec/R-REC-P.2001-4-202109-I/en (accessed on 10 April 2022).
26. Recommendation ITU-R P.618, Propagation Data and Prediction Methods Required for the Design of Earth-Space Telecommunication Systems. 2017. Available online: https://www.itu.int/rec/R-REC-P.618-13-201712-I/en (accessed on 21 April 2021).
27. Recommendation ITU-R P.679-4, Propagation Data Required for the Design of Broadcasting-Satellite Systems. 2015. Available online: https://www.itu.int/\rec/R-REC-P.679-4-201507-I/en (accessed on 21 April 2021).
28. Recommendation ITU-R P.680, Propagation Data required for the Design of Earth-Space Maritime Mobile Telecommunication Systems. 2022. Available online: https://www.itu.int/rec/R-REC-P.680-4-202208-I/en (accessed on 9 November 2022).
29. Recommendation ITU-R P.681, Propagation Data Required for the Design Systems in the Land Mobile-Satellite Service. 2019. Available online: https://www.itu.int/rec/R-REC-P.681-11-201908-I/en (accessed on 21 April 2021).
30. Recommendation ITU-R P.682, Propagation Data Required for The Design of Earth-Space Aeronautical Mobile Telecommunication Systems. 2022. Available online: https://www.itu.int/rec/R-REC-P.682-4-202208-I/en (accessed on 9 November 2022).
31. Han, H.; Yang, C.; Li, J.; Wang, J.; Wang, J. Research on the Applicability of ITU-R P. 2001 in the Typical Area of China. *J. CAEIT* **2022**, *17*, 977–1006.
32. Recommendation ITU-R M.1583-1, Interference Calculations Between Non-Geostationary Mobile-Satellite Service or Radionavigation-Satellite Service Systems and Radio Astronomy Telescope Sites. 2007. Available online: https://www.itu.int/rec/R-REC-M.1583/en (accessed on 21 April 2021).

Disclaimer/Publisher's Note: The statements, opinions and data contained in all publications are solely those of the individual author(s) and contributor(s) and not of MDPI and/or the editor(s). MDPI and/or the editor(s) disclaim responsibility for any injury to people or property resulting from any ideas, methods, instructions or products referred to in the content.

universe MDPI

Article

A Meteor Detection Algorithm for GWAC System

Yicong Chen [1], Guangwei Li [2,*], Cuixiang Liu [1,*], Bo Qiu [1,*], Qianqian Shan [1] and Mengyao Li [1]

1 School of Electronic and Information Engineering, Hebei University of Technology, Tianjin 300401, China; 202121902006@stu.hebut.edu.cn (Y.C.); 202221902012@stu.hebut.edu.cn (Q.S.); 202221902026@stu.hebut.edu.cn (M.L.)
2 Key Laboratory of Space Astronomy and Technology, National Astronomical Observatories, Chinese Academy of Sciences, Beijing 100101, China
* Correspondence: lgw@bao.ac.cn (G.L.); liucuix@hebut.edu.cn (C.L.); qiubo@hebut.edu.cn (B.Q.)

Abstract: Compared with the international meteor surveillance systems, the ground wide angle camera (GWAC) system exhibits characteristics such as images with the resolution of 4K × 4K and single-site observation. These characteristics present challenges for meteor detection in the GWAC system. Consequently, this paper proposes a new meteor detection algorithm for the GWAC system on the base of the solely mini-GWAC system data algorithm. The new algorithm consists of the following key steps: (1) to compare differences between adjacent frames, applying block-based image binarization thresholds, and incorporating median filtering to reduce noise; (2) to adopt the probabilistic Hough transform (PHT) to identify moving objects and cluster them based on the origin moment of the line segments, while assessing the credibility of clustering; (3) to introduce the so-called maximum disappearance frame for moving objects in the tracking algorithm, enhancing the ability to track multi-frame moving objects. The utilization of the line segment inclination angle of the moving object as the direction of movement facilitates the tracking of multiple moving objects, thereby reducing the probability of mistakenly selecting single-frame moving objects; (4) to leverage the light curves of single-frame moving objects to select meteors to enhance the accuracy of meteor detection. Comparative experiments demonstrate that our proposed algorithm processes each frame image in just 0.39 s, achieving an accuracy of 89.8% in the dataset of 5856 adjacent frames. The experimental results indicate that the algorithm achieved an accuracy of 90.27% when applied in the meteor detection of the image data captured by the GWAC system from Dec. 10th to 19th in 2019 and 2021, obtaining excellent detection results.

Keywords: meteor detection; GWAC; moving objects tracking; light curve

1. Introduction

The origin of meteor astronomy can be traced back to the 19th century [1–3], when scientists began to systematically study and observe meteors. For example, understanding the characteristics of meteors can provide information about the interplanetary dust environment surrounding Earth and how it has evolved during the solar system's evolutionary process. Meteor automation processing is a critical technology that enables efficient detection, tracking, and analysis of meteors. Scientists can obtain accurate meteor trajectories and mass characteristics, thereby promoting research on the origin and physical properties of meteors [4].

Most of the existing international meteor networks detect meteors through temporal and spatial correlations [5], primarily utilizing techniques such as temporal difference, pixel averaging, thresholding, and Hough transform (HT). MeteorScan [6] compares frame differences between adjacent time image pairs and performs HT recognition and matched filtering to reduce false detection probability. MetRec [7] reduces noise by averaging pixels in a 2 × 2 fashion and downsamples the image size by a factor of four. It employs multi-frame region of interest (ROI) tracking to detect meteor, but this results in higher computational complexity. In contrast, UFOcapture [8] does not utilize mean differencing and spatial-temporal correlation. It applies a 5 × 5 spatial filter with frame differencing that is then masked and thresholded. ASGARD [9,10] performs real-time detection by comparing the pixels of the current video frame with the previous 10 frames, counting the number of pixels with increased brightness that exceed a set threshold to reduce noise interference.

There are a number of meteor surveillance systems internationally. International Meteor Organization (IMO) [11] Video Meteor Network, established by Germany, currently consists of 88 cameras operated by 49 observers from 16 countries. They use MetRec to capture meteor tracks from video sequences at the standard frame rates of 25 or 30 FPS of the cameras. SonotaCo Network, located in Japan, consists of 70 observation stations. It utilizes stars for field calibration and employs software such as UFOCapture, UFOAnalyzer and UFOOrbit to detect and analyze meteors. The Cameras for All-sky Meteor Surveillance (CAMS) network [12] consists of three observation stations, with each station operating 60 cameras. It can monitor the sky at altitudes above 31 degrees. MeteorScan-based trajectory search technology autonomously processes meteor images acquired from each station. European viDeo MeteOr Network Database (EDMOND) [13] is generated through collaboration and data sharing among several European national networks and IMO Video Meteor networks. It collects data from 155 sites and utilizes UFOOrbit for automated processing of meteors. The nine-channel Mini-Mega TORTORA (MMT-9) [14] was deployed in 2014 at a specialized astrophysical observatory near the 6 m Russian telescope. It offers high time-resolution with exposure times ranging from 0.1 to several hundred seconds. It has the capability to observe a wide range of 900 square degrees or a narrow range of 100 square degrees of the sky simultaneously, enabling real-time data processing and effective detection of various transient events. This system has detected over 90,000 meteors. Meteors occur at a rate of 300–350 per night, with durations ranging from 0.1 to 2.5 s and angular velocities reaching up to 38 degrees per second. The Canadian Automated Meteor Observatory (CAMO) [15] is an automated video meteor system comprising two sites. It comprises two enhanced cameras: a wide-field camera with a 28° field of view, utilized for collecting meteor light curves and calculating trajectories; and a narrow-field camera with a 1.5° field of view, employed for tracking meteors and obtaining high-resolution observations. The cameras have a resolution of 640 × 480, with the wide-field camera capturing 80 frames per second and the narrow-field camera capturing 110 frames per second. ASGARD is used for real-time detection of the wide-field images to guide the narrow-field tracking of meteors.

GWAC [16–18], established at the Xinglong Observatory of the National Astronomical Observatories of China, is used for the measurement of optical transient objects before, during, and after gamma-ray bursts [19]. It consists of 40 joint field of view (JFoV) cameras. Each camera, with an aperture of 18 cm and custom-made lenses with an f-ratio of f/1.2, is equipped with a 4k × 4k E2V back-illuminated CCD chip. The wavelength range is from 0.5 to 0.85 μm. The field of view for each camera is 150 $\deg^2$ and the pixel scale is 11″7. The total field of view for each unit carrying four JFoV cameras is ~600 $\deg^2$. The entire camera array covers an area of 5000 $\deg^2$ in the sky, capturing an image every 15 s, including a 10-s exposure and 5-s readout [20]. The limiting magnitude reaches a V-magnitude of 16, allowing for the detection of various moving objects such as meteors, asteroids, comets, airplanes, space debris, and satellites. The maximum brightness value before the system reaches saturation is 8 mag. Continuous observations are conducted each observing night for up to 10 h on specific areas of the sky. The GWAC system is designed for short-timescale sky surveys, aiming to detect accurately and rapidly astronomical anomalies. The detection of meteors is a significant component of the GWAC system, offering valuable insights into meteor trajectories, brightness levels, and mass distribution information.

International meteor surveillance systems are characterized by multi-site observations, primarily using the video, and focusing on brighter meteors [21]. However, the GWAC system differs in the following aspects: (1) It lacks the ability to measure the height and velocity information of moving objects through single-site observations. (2) It captures images at a frequency of every 15 s, with a resolution of 4k × 4k. While meteor duration typically ranges from 0.1 to 2.5 s, so it often appears as single frames in GWAC system. (3) There is a presence of a significant amount of background moving objects. The current international meteor detection algorithms are not applicable for the meteor detection in the GWAC system. Xu Yang and colleagues [22] conducted a study on identifying meteor candidates using two months of data from the mini-GWAC system in 2019. The mini-GWAC system has a lower image resolution of only 3k × 3k and captures fewer background stars and moving objects. They provide data indicating that aircraft can achieve angular velocities of 1.85 degrees per second and satellites can attain angular velocities of 0.025 degrees per second with durations of 10 s. This enables them to remain within the image frame for an extended period before exiting. As a result, they exhibit continuous presence across multiple frames. We exclude non-meteor objects by considering their presence in multiple frames. Consequently, using this algorithm for meteor candidate detection for the GWAC system resulted in several issues, including missing and false detection of moving objects, incomplete clustering of moving objects, and low precision in single-frame moving-object filtering. Therefore, we presented a meteor detection algorithm specifically designed for the GWAC system based on the algorithm proposed in Reference [22]. Our algorithm can detect meteors from GWAC images with a magnitude range of about −0.66~−7.26 mag. For the brightest meteor, the GWAC can only record a part of it, so it should be brighter than the magnitude given by the GWAC. Our proposed algorithm aims to achieve real-time meteor detection while improving accuracy.

The main contributions of this study are as follows: (1) to apply image block division to set image binarization threshold and introduce median filtering to reduce noise and detect more line segments, avoiding the issues of false detection and missed detection of numerous moving objects. (2) to utilize the origin moment information of the line segment in the moving objects clustering model, and we employ the line segment inclination angle information to verify the credibility of the clustering objects to improve the accuracy of moving objects clustering. (3) to propose a moving object tracking algorithm. The concept of the maximum disappearance frame for moving objects introduced in the moving objects tracking algorithm contributes to tracking multi-frame moving objects. The inclination angle of the line segments of moving objects utilized as the direction of movement facilitates the tracking of multiple moving objects. Thereby it can improve the probability of correctly filtering single-frame moving objects and the accuracy of meteor detection. (4) to conduct comparative experiments between our proposed algorithm and the Reference [22] algorithm to validate the effectiveness. Our algorithm achieves a precision of 89.4% and an improvement of 8.4%, resulting in an average processing time of 0.39 s per image frame. Additionally, we utilized data collected by GWAC system from 10th to 19th December in 2019 and 2021 for meteor detection. The precision of our algorithm was found to reach 90.27%.

2. Materials and Methods

This study utilizes data collected from the GWAC system and proposes a new meteor detection algorithm based on the mini-GWAC system algorithm. In the first step, we apply the temporal difference method to remove non-transient objects from adjacent frame images. Subsequently, we employ the PHT to identify the line segments corresponding to the detected moving objects after conducting image preprocessing. A moving objects clustering algorithm is applied in order to cluster these line segments. In the second step, we implement a single-frame moving object tracking algorithm that utilizes the angle feature of the line segments to effectively filter single-frame moving objects. Finally, by analyzing the light curves of the single-frame moving objects, we can classify the single- and double-peaked single-frame moving objects as potential meteor candidates. The flowchart illustrating our meteor detection algorithm is presented in Figure 1.

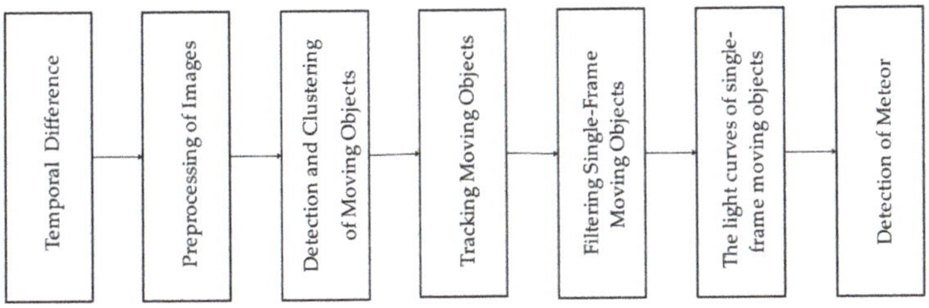

Figure 1. Flowchart of our meteor detection algorithm.

2.1. Dataset

We collected data from the GWAC system during the Gemini meteor shower period, including seven days from 12 December to 18 December 2021, and eight days from 10 December to 14 December and 17 December to 19 December 2019. The collected data for each date are shown in Figure 2.

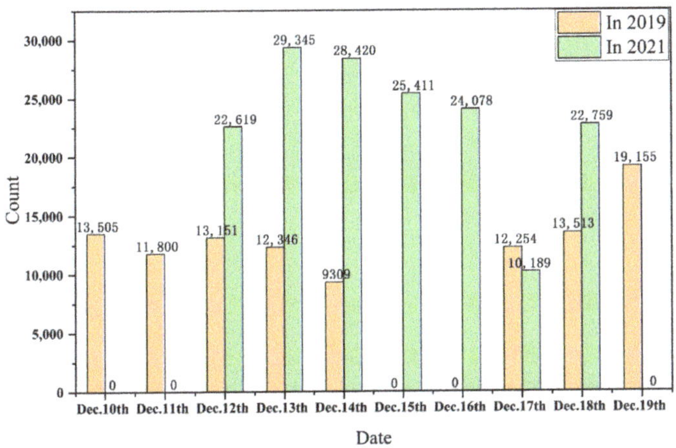

Figure 2. Raw events of experimental data for each date in 2019 and 2021.

The GWAC system has a time sampling period of 15 s, and each camera can capture a maximum of 240 images per hour. It captures the fixed sky area for approximately 10 h during each observing night. From 10 December to 19 December 2019, a total of 105,033 images were collected by the GWAC system. Similarly, the GWAC system captured a total of 162,821 images from 12 December to 18 December 2021. Consequently, the total number of images to test obtained in 2019 and 2021 was 267,854.

2.2. Meteor Detection Model

2.2.1. Temporal Difference

Temporal difference [23] is a widely employed technique for detecting moving objects, applied diverse applicability in various domains including video analysis, video surveillance, and action recognition. It detects moving objects by subtracting the pixel values between adjacent frames.

$$D_t(x,y) = I_t(x,y) - I_{t-1}(x,y) \tag{1}$$

where $I_{t-1}(x,y)$ represents the pixel values of the $t-1$ frame image, $I_t(x,y)$ represents the pixel values of the t frame image, and $D_t(x,y)$ represents the pixel differences between the adjacent and $t-1$ frame images. In order to identify transient moving objects, it is used to remove stars, filter out sporadic noise and cosmic rays to reduce the influence of background moving objects.

2.2.2. Preprocessing of Images

Preprocessing operations need to be applied to the differenced images before recognizing moving objects. Image preprocessing includes image binarization and median filtering. Firstly, the differenced images are subjected to binary thresholding. The images captured by the GWAC system have a high resolution of 4k × 4k. The binary thresholding is set as twice the standard deviation of the average pixel across the entire image. There is a risk of losing potential moving objects if the threshold is set too high in cases of bright backgrounds. Likewise, setting the threshold too low in cases of dark backgrounds can result in substantial background noise. Therefore, it is divided into 2 × 2 small images, resulting in four regions $d_i(i = 1,2,3,4)$. The threshold T_i for each region is calculated as $T_i = \mu_i + 2\sigma_i(i = 1,2,3,4)$, where μ_i and $\sigma_i{}^2$ represent the average pixel intensity and variance of each region, respectively. The formula for image binarization is as follows:

$$d_i(x,y) = \begin{cases} 255 & d_i(x,y) > T_i \\ 0 & \text{others} \end{cases} (i = 1,2,3,4) \tag{2}$$

Pixels above the threshold are assigned the color white, while pixels at or below the threshold are assigned the color black. Utilizing an adaptive block-wise adjustment of the binary thresholding allows for a more precise differentiation between moving objects and their background, resulting in reduced background noise and accurate preservation of the detailed trajectories of the moving objects. Subsequently, the binary images undergo median filtering to reduce noise and prevent interference with PHT for detection. Recognizing that moving objects usually cover multiple pixels, a 3 × 3 template is deliberately chosen for median filtering to effectively remove isolated noise points.

2.2.3. Detection and Clustering of Moving Objects

The differenced images are utilized to distinguish the moving objects from the background following image preprocessing. The moving objects appear as straight lines, which can be recognized using HT. Hough transform [24] is based on the concept of mapping a pixel point (x,y) from the image space to a sinusoidal curve $r = x\cos\theta + y\sin\theta$ of the polar coordinate space (r,θ). The sinusoidal curve represents all possible straight lines passing through that point. Collinear points in the image space are represented by a single point (r',θ') at which sinusoidal curves in the polar coordinate space intersect. Straight lines can be detected by evaluating the accumulation value at that point (r',θ') in the polar coordinate space, as illustrated in Figure 3. HT can also be applied to detect geometric shapes such as circles, ellipses, and other specific shapes in addition to line detection [25,26].

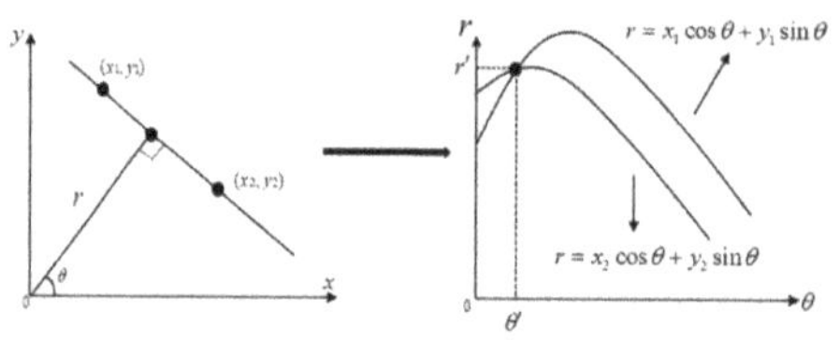

Figure 3. Points in the image space to sinusoidal curves in the polar coordinate space.

HT exhibits high computational complexity, particularly when handling large-sized images. The PHT is employed as a solution to mitigate this problem by reducing the computational burden. A subset of randomly selected points from the image is considered as sample points in the PHT, which then finds other connected sample points and accumulates votes based on their relationships to determine the parameters of the lines. The linear features and positions of the moving object can be extracted by utilizing the computation and analysis of the PHT [27].

The PHT recognizes moving objects, generating multiple line segments forming a set. Hence, the clustering of line segments from the moving objects is an essential step. The process of clustering line segments utilizes two primary features [28]: the line segment origin moment d and the inclination angle θ. The implementation method of line segments clustering is as follows:

1. The new clustering object set is initialized by adding the origin moment d, the average origin moment $\overline{d}$, two position coordinates (x_1, y_1) and (x_2, y_2), the inclination angle θ, the average of the inclination angle $\overline{\theta}$ and the variance of the inclination angle $\hat{\theta}$ from the first line segment or unmatched line segment in the line segments set.

$$\Delta y = y_2 - y_1, \Delta x = x_2 - x_1 \tag{3}$$

$$\theta = \arctan\frac{\Delta y}{\Delta x} \tag{4}$$

$$l = \sqrt{(\Delta x)^2 + (\Delta y)^2} \tag{5}$$

$$d = \frac{|y_1 \Delta x - x_1 \Delta y|}{l} \tag{6}$$

2. The distance is calculated between the origin moment d of the line segment in the line segment set and the average origin moment $\overline{d}$ of each clustering object set in turn. If the distance is less than the maximum distance error (MDE), the line segment is classified into the same clustering object set, $\overline{d}$, $\overline{\theta}$ and $\hat{\theta}$ are updated in clustering object set. If the line segment is matched with all clustering object sets, step 1 is executed to add a new clustering object set.
3. When all the line segments in the line segment set are clustered, the clustering reliability of each clustering object set is judged by $\hat{\theta}$ of the line segments within each clustering object set.
4. If the variance of the inclination angles $\hat{\theta}$ of the clustering object set is less than the maximum inclination angle error (MIAE), it indicates a successful clustering set; otherwise, it fails. Then the inclination angle error is calculated between the inclination angle θ of the line segments of the failed clustering object set and the average inclination angle $\overline{\theta}$ of each clustering object set in turn. If it is less than MIAE, this line segment is matched with this clustering object set; otherwise, step 1 is executed to add a new cluster object set.
5. Each clustering object set represents a moving object. The longest line segment in the clustering object set is denoted as the longest trajectory of the moving object, with position coordinates (x_{min}, y_{min}) and (x_{max}, y_{max}).

2.2.4. Tracking Moving Objects and Filtering Single-Frame Moving Objects

Meteor characteristics in the GWAC system primarily appear in single frames, while non-meteor objects tend to appear in multiple frames. We track moving objects in each frame to detect if the moving objects in the current frame match with the previous ones, consequently filtering single-frame moving objects [29]. The main implementation method is as follows:

6. The position coordinates (x_{min}, y_{min}) and (x_{max}, y_{max}), the inclination angle θ, and midpoint coordinate (x_{mid}, y_{mid}) of each moving object in every frame are obtained. If it is the first frame, each moving object is assigned an initial ID.
7. It is firstly necessary to determine whether the moving objects in the current frame match with the ones in the previous frames.

8. We calculate the inclination angle θ' of the line segments formed by pairing the midpoints (x_{mid}, y_{mid}) of the moving objects in the current frame with the midpoints (x_{obj}, y_{obj}) of the moving objects in the tracking object set, and create a matrix M of the inclination angles θ'.

$$M(i,j) = \arctan\frac{y_{mid}(i) - y_{obj}(j)}{x_{mid}(i) - x_{obj}(j)}(i = 1,2,3,\ldots,m; j = 1,2,3,\ldots,n) \quad (7)$$

9. Where the total of moving objects in the current frame is m, while the tracking object set consists of n moving objects.
10. We update the ID and the position coordinate of the moving objects in the current frame. The inclination angle θ_j of the moving object j in the tracking object set indicates its movement direction in the next frame. The inclination angles $\theta'_{i,j}$ in column j of the matrix M indicate the actual movement directions of the moving object j. Then we find the moving object i in the current frame that inclination angle $\theta'_{i,j}$ is the nearest θ_j and calculate the error in the inclination angle between the two. If this error is less than MIAE, the moving object i in the current frame is successfully matched the moving object j in the tracking object set. They are regarded as the same object and assigned the same ID. The appearance count of this moving object j is added by 1.
11. If there are moving objects matched unsuccessfully in the tracking object set, they are marked as disappeared moving objects. When the number of frames in which they have disappeared exceeds the maximum disappearance frame (MDF), they are removed from the tracking object set.
12. If there are moving objects matched unsuccessfully in the current frame, they are marked as newly appeared moving objects and assigned a new ID. They are added to the tracking object set along with their position coordinates and inclination angles. The appearance count of these objects is incremented by 1.
13. We filter single-frame moving objects. When the moving object is marked as disappeared and the number of frames it has been missing exceeds the MDF, it is considered a single-frame moving object if the appearance count of it is 1.

In the moving-object tracking algorithm, the concept of the MDF for moving objects introduced in the moving objects tracking algorithm is contributed to track multi-frame moving objects. The inclination angle of the line segments of moving objects utilized as the direction of movement facilitates the tracking of multiple moving objects. Thereby it can reduce the probability of incorrectly filtering single-frame moving objects and improve the accuracy of meteor detection.

2.2.5. Meteor Detection

In addition to utilizing the meteor characteristics in the GWAC system appearing in a single frame, meteor features can also be used for further filtering single-frame moving objects to select meteor candidates and improve the accuracy of meteor detection. Meteors typically exhibit a brief bright light and a long tail, distinguishing them from other moving objects such as airplanes and satellites, as shown in Figure 4.

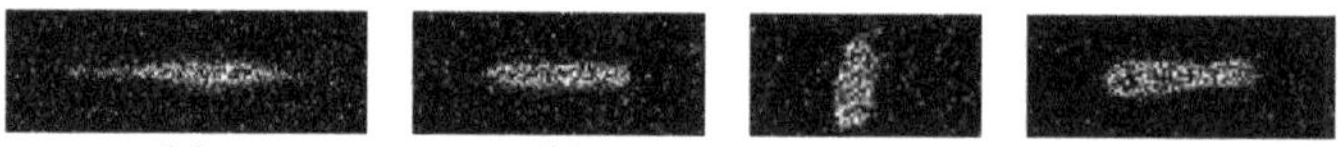

(a) **(b)** **(c)** **(d)**

Figure 4. Different moving objects observed for the GWAC system: (**a**) meteor; (**b**–**d**) other moving objects.

The variation of brightness of celestial bodies over time is referred to as the light curve in astronomy [30]. It is not possible to obtain the variation curve of brightness of moving objects over time in the GWAC system. Therefore, the variation curve of brightness of moving objects with respect to the image pixels is used for the light curve and defined as the trajectory of the moving objects [22]. By calculating the light curves of single-frame moving objects, we analyze their shapes, and filter out non-meteor single-frame moving objects to obtain meteor.

- Preprocessing of single-frame moving objects

The images are rotated clockwise α to obtain horizontally moving objects. The rotation transformation is as follows:

$$(x'\ y'\ 1) = (x\ y\ 1)\begin{pmatrix} 1 & 0 & 0 \\ 0 & -1 & 1 \\ -0.5w & 0.5h & 1 \end{pmatrix}\begin{pmatrix} \cos\alpha & -\sin\alpha & 0 \\ \sin\alpha & \cos\alpha & 0 \\ 0 & 0 & 1 \end{pmatrix}\begin{pmatrix} 1 & 0 & 0 \\ 0 & -1 & 1 \\ 0.5w' & 0.5h' & 1 \end{pmatrix} \quad (8)$$

where the co-ordinates of the image before rotation are (x, y), and after rotation they are (x', y'). w and h represent the width and height of the image before rotation. w' and h' represent the width and height of the image after rotation. The rotation angle α is $180° - \theta$. The position co-ordinates of the single-frame moving object before rotation are (x_{min}, y_{min}), (x_{max}, y_{max}), and after rotation they are (x'_{min}, y'_{min}) and (x'_{max}, y'_{max}). The rotation angle of the moving object is θ. Next, the region is cropped, and the cropping area is as follows.

$$area(x, y) = \begin{bmatrix} (x'_{min} - 30,\ y'_{min} - 10) & \cdots & (x'_{\max} + 30,\ y'_{min} - 10) \\ \vdots & \ddots & \vdots \\ (x'_{min} - 30,\ y'_{min} + 10) & \cdots & (x'_{min} - 30,\ y'_{min} + 10) \end{bmatrix} \quad (9)$$

- The light curves of single-frame moving objects

We sum the brightness along the vertical axis with respect to the horizontal axis in the cropping area of the single frame moving object, resulting in the light curve.

$$F(x) = \sum\nolimits_y area(x, y) \quad (10)$$

The light curve exhibits abundant spike noise. To reduce the noise, we perform a smoothing convolution, resulting in smoothed light curves. We then detect the peaks in the light curves and classify them into flat peaks, single-peaked, or double-peaked moving objects based on the number of peaks. This is illustrated in Figure 5.

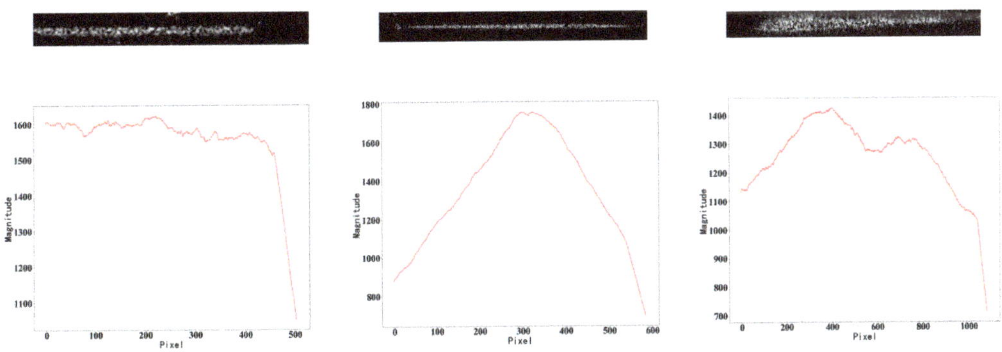

(a) (b) (c)

Figure 5. The light curves of single-frame moving objects in the GWAC system: (**a**) A flat peak object and its light curve; (**b**) A single-peaked object and its light curve; (**c**) A double-peaked object and its light curve.

According to [30], flat curves of meteors only account for 15% of the total classified objects in meteor events. The category of flat curves comprises a small number of meteors but large number of non-meteor objects. To obtain highly accurate meteor candidates, single-peaked and double-peaked moving objects are selected, filtering out non-meteor moving objects.

3. Results and Discussions

3.1. Implementation Details

The meteor detection algorithm was implemented using Python and OpenCV. The experiments were conducted on an AMD Ryzen 7 4800H CPU and a NVIDIA GeForce GTX 1650 Ti GPU.

The parameters for PHT are established as follows: ρ is set to one pixel, the accumulator threshold parameter is set to 35, the minimum length of line segments is set to 35 pixels, and the maximum allowed gap between line segments is set to 10 pixels. As for line-segment clustering, the maximum distance error is limited to 200 pixels, and the maximum inclination angle error is constrained at 3°. As for moving objects tracking, it is ensured that the maximum disappearance frame are kept at 3.

The results of the object detection are evaluated using precision (P) and recall (R). Precision represents the probability of correctly identifying meteor samples among the data identified as meteor samples. Recall represents the probability of correctly identifying meteor samples among meteor samples in the dataset.

$$\mathrm{P} = \frac{TP}{TP + FP} \tag{11}$$

$$\mathrm{R} = \frac{TP}{TP + FN} \tag{12}$$

where *TP* represents true positive, *TN* represents true negative, *FP* represents false positive, and *FN* represents false negative.

3.2. Experimental Results

3.2.1. Comparative Experiments

To provide a visual and accurate comparison between our proposed algorithm and the algorithm proposed in Reference [22], Figure 6 displays the results of temporal difference and image preprocessing for both algorithms. From the comparison shown in Figure 6a,b, moving objects are not observed. Figure 6c shows the presence of overexposed regions after temporal difference of the Reference [22] algorithm. In Figure 6d, image preprocessing does not handle the regions, resulting in false detections in the PHT. In contrast, it can be observed that our proposed algorithm does not have overexposed regions after the temporal difference and image preprocessing from Figure 6e,f. The noise is significantly reduced, avoiding false detection of moving objects.

The results of image preprocessing and PHT for both the Reference [22] algorithm and our proposed algorithm are showed in Figure 7. The comparison between Figure 7a,b reveals the presence of transient moving objects in adjacent frames. After applying the image preprocessing of the Reference [22] algorithm, a significant amount of noise interferes with the subsequent recognition for the PHT in Figure 7c, leading to the absence of detected moving objects in Figure 7d. However, the image preprocessing with block-based thresholding and median filtering significantly reduces noise in Figure 7f, resulting in the detection of moving objects in Figure 7e using the PHT. Therefore, the comparative experiments demonstrate that block-based thresholding for image binarization and median filtering can effectively reduce noise and prevent the occurrence of missed detection of moving objects.

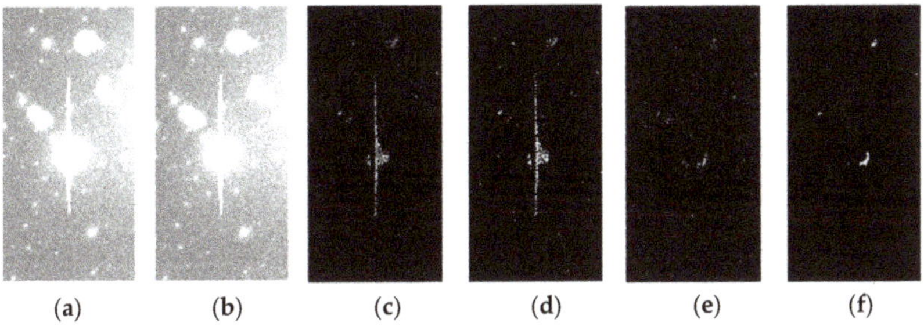

Figure 6. The results of the adjacent frame temporal difference and image preprocessing: (**a**) A partial view of the G024_Mon_objt_211214T20201965 (frame t); (**b**) A partial view of the G024_Mon_objt_211214T20203464 (frame t + 1); (**c**,**d**) The results of the Reference [22] algorithm; (**e**,**f**) The results of our proposed algorithm.

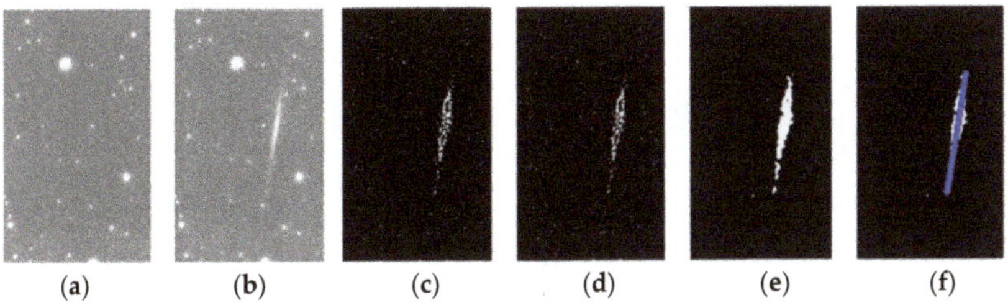

Figure 7. The results of the adjacent frame image preprocessing and PHT: (**a**) A partial view of the G024_Mon_objt_211214T19104960 (frame t); (**b**) A partial view of the G024_Mon_objt_211214T19110460 (frame t + 1); (**c**,**d**) The results of the Reference [22] algorithm; (**e**,**f**) The results of our proposed algorithm.

The results of the line segments clustering for both the Reference [22] algorithm and our proposed algorithm are displayed in Figure 8. From Figure 8a, it can be observed that there is only one moving object in the image G024_Mon_objt_211214T20281960. However, Figure 8b and its bottom-left magnified portion show that the Reference [22] algorithm clusters this moving object into two separate ones. Figure 8c demonstrates the effectiveness of our proposed algorithm in clustering the moving object into a single object.

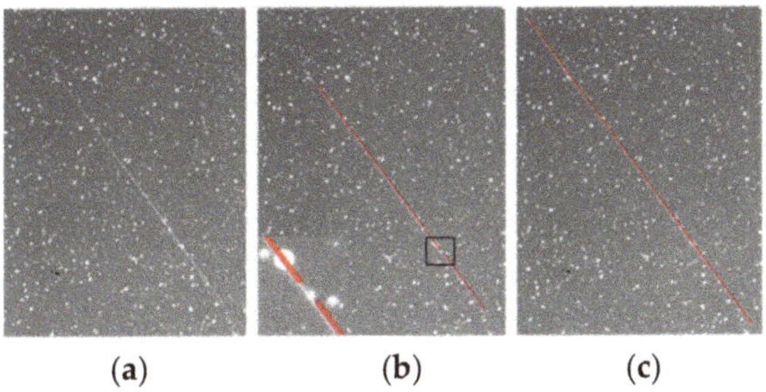

Figure 8. The results of line segments clustering: (**a**) A partial view of the G024_Mon_objt_211214T20281960; (**b**) The clustered line segments by the Reference [22] algorithm; (**c**) The clustered line segments by our proposed algorithm. **Note:** The red lines represent the clustering results, where each line corresponds to a clustering moving object.

The Reference [22] algorithm only considers the tracking of adjacent frames for moving objects. When a moving object appears in adjacent frames and is missed in a certain frame, there is possibility of mistakenly identifying the last frame's moving object as a single-frame moving object. As shown in Figure 9a–c, when a moving object appears in three adjacent frames, it is labeled as ID:0 in the first frame. If this moving object is not detected in the second frame, resulting in the failed match, the first frame of this moving object is erroneously classified as a single-frame moving object. This, in turn, affects the accuracy of meteor detection. Our proposed algorithm sets the MDF as three times, meaning that object can still be matched with the same object even after disappearing for three frames. As shown in Figure 9d–f, when a moving object appears in the first frame with ID:0 and is matched with the moving object in the second frame, if the second frame doesn't detect this moving object resulting in a failed match, the position co-ordinates of the moving object from the first frame are retained, and it is considered as one disappearance of the object. Later, when matched with the moving object in the third frame, it is labeled as ID:0, accurately identifying the single-frame moving object. Since it reappears several frames later, we can exclude that this was a meteor.

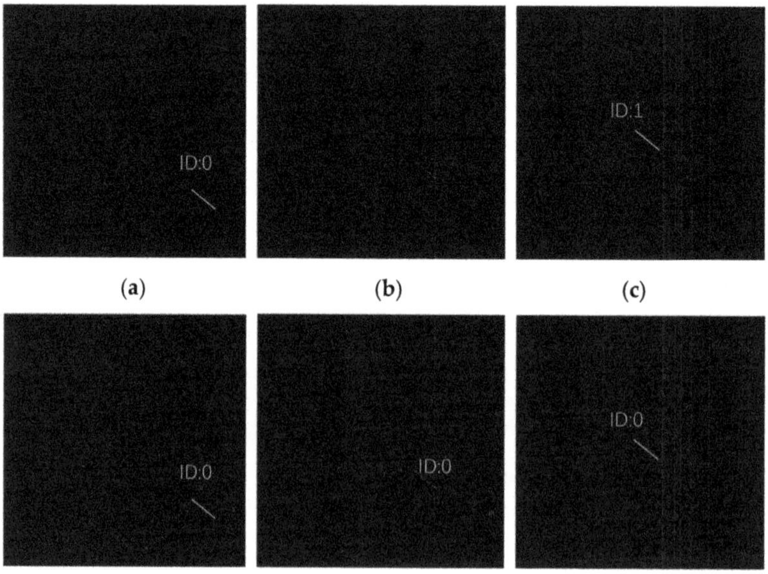

Figure 9. The tracking results of multi-frame moving object: (**a**–**c**) The tracking results of multi-frame moving object by the Reference [22] algorithm; (**d**–**f**) The tracking results of multi-frame moving object by our proposed algorithm.

The Reference [22] algorithm only utilizes the inclination angle of adjacent line segments, which leads to the problem of matching the last frame's moving objects with multiple moving objects in the current frame as the same object. As shown in Figure 10a,b, the inclination angle of the moving object in the first frame is 154.18°, while in the second frame it is 153.86°. The inclination angle of the new moving object (left side in Figure 10b or Figure 10d) is 156.98°. The Reference [22] algorithm assigns ID:0 to the moving object in the first frame, and all moving objects in the second frame that have the inclination angle within 5° difference from the first frame's moving object are matched, resulting in mistakenly identifying the newly appeared moving object in Figure 10b as the same moving object. Our proposed algorithm, on the other hand, utilizes the inclination angle of the line segments from the first frame's moving object as the reference for the movement direction of the next frame's moving object. The inclination angle between the midpoint of the second frame's (current frame) moving object and the midpoint of the previous frame's moving objects is considered as the direction of the actual trajectory. If the difference in the direction inclination angle is less than 3°, they are matched as the same object. As shown in Figure 10d, the moving object in the second frame is successfully matched with the moving object in the first frame, with ID:0, and a new moving object with ID:1 is detected.

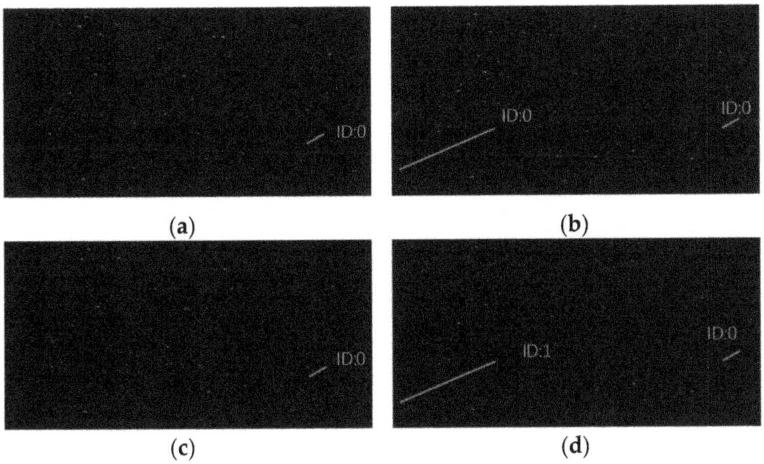

Figure 10. The tracking results of multiple moving objects: (**a**,**b**) The tracking results of multiple moving objects by the Reference [22] algorithm; (**c**,**d**) The tracking results of multiple moving objects by our proposed algorithm.

The visual results of the experimental comparison between our proposed algorithm and the Reference [22] algorithm from Figures 6–10 demonstrate our proposed algorithm can effectively reduce noise, prevent missed detections of moving objects, accurately identify the single-frame moving object and track multiple moving objects. This validates the effectiveness of our algorithm.

In this study, 5856 adjacent frames images from 14 December 2021 were selected as the dataset for accurate comparison. Single-frame moving objects were filtered and further refined to select meteor candidates. The selected dataset was manually annotated, revealing a total of 242 single-frame moving objects and 189 meteor candidates. The algorithms proposed by Reference [22] and this study were used to filter the single-frame moving objects, and precision and recall were calculated for each algorithm. In addition, there are a few edge objects in the single-frame moving objects. After removing the edge objects, the Reference [22] algorithm and our proposed algorithm were used to detect meteor candidates. Similarly, precision and recall were calculated. The calculation results are shown in Table 1. The precision of our proposed algorithm in single-frame moving object selection reaches 91.4%, which is an improvement of 13.3% compared to the Reference [22] algorithm. Out of 242 single-frame moving objects, 88.0% were correctly filtered. In terms of meteor selection, our algorithm achieves a precision of 89.4%, showing an improvement of 8.4% compared to the Reference [22] algorithm. Out of 189 meteor objects, 83.5% were accurately detected. The high precision of the meteor detection algorithm ensures the high probability of correctly identifying meteor samples among the data identified as meteor samples, which provides valuable positional information for further research on the brightness and mass of meteors.

Table 1. The precision and recall for filtering single-frame moving objects and detection of meteors by the Reference [22] algorithm and our proposed algorithm.

Method	Filtering Single-Frame Moving Objects		Detection of Meteors	
	Precision	Recall	Precision	Recall
The Reference [22]	0.781	0.759	0.814	0.793
Our algorithm	0.914	0.880	0.898	0.835

The time characteristic of the entire meteor detection process was evaluated by comparing the Reference [22] algorithm with our proposed algorithm on the dataset of 5856 image frames. The summarized experimental results can be found in Table 2. The Reference [22] algorithm required 6965 s to process the 5856 image frames, resulting in an average processing time of 1.19 s per image frame. In contrast, our proposed algorithm completed the process in 2280 s, resulting in an average processing time of 0.39 s per image frame. This indicates an average improvement of 0.8 s per image frame compared to the Reference [22] algorithm, satisfying the real-time requirements of the GWAC for meteor detection.

Table 2. The time of the entire meteor detection process by the Reference [22] algorithm and our proposed algorithm.

Method	The Reference [22]	Our Algorithm
Time/s	6965	2280

3.2.2. Meteor Detection for the GWAC Data

Firstly, moving-object identification was conducted on 267,854 images from 2019 and 2021. The results showed that 11,460 moving objects in 2019 and 26,523 moving objects in 2021, totaling 37,983 moving objects, were identified. Subsequently, the identification of meteor objects required further selection of single-frame moving objects. The filtering results revealed that 2759 single-frame moving objects were selected in 2019 and 6742 single-frame moving objects were selected in 2021, with a total of 9501 single-frame moving objects. In 2019, single-frame moving objects accounted for 24.08% of the total number of moving objects, while in 2021 they accounted for 25.41%. The overall selection of single-frame moving objects in 2019 and 2021 accounted for only 24.75% of the total number of identified moving objects, indicating that non-meteor objects such as airplanes and satellites with multiple frames constituted a higher proportion of the total moving objects. The results of total identified moving objects and single-frame moving objects for each date in 2019 and 2021 are presented in Figure 11. On 13 December 2021, during the peak period of the Geminid meteor shower, the highest number of single-frame moving objects was identified, reaching up to 1350. Near the peak period on 14 December 2021, 1224 single-frame moving objects were identified.

Finally, meteor detection was performed on the single-frame moving objects from 2019 and 2021, obtaining highly accurate meteor candidates. The detection of meteors for each date in 2019 and 2021 is presented in the following Table 3.

There were 11.28% edge-moving objects detected among the selected single-frame moving objects in 2019 and 2021. These edge-moving objects only represent partial trajectories, making them indeterminate, and therefore classified as flat-peak moving objects. In 2019, a total of 1719 meteor candidates were selected, out of which 1544 were confirmed through manual verification, resulting in an accuracy of 89.82%. Similarly, 4295 meteor candidates were selected in 2021, with 3897 confirmed through manual verification by checking their light curves, resulting in an accuracy of 90.73%. The average precision in meteors detection for 2019 and 2021 reached 90.27%, with meteor candidates accounting for 15.83% of the total moving objects. From Table 3, it is evident that on December 13th in

2019 and 2021, the Gemini meteor shower reached its peak, with the highest number of detected meteors.

Table 3. Number of single-frame moving objects and meteors for each date in 2019 and 2021.

Date	Number of Single-Frame Moving Objects (2019/2021)	Number of Meteors (2019/2021)	Manually Verified Number of Meteors (2019/2021)
10 December	358/0	208/0	179/0
11 December	331/0	228/0	197/0
12 December	370/938	216/621	201/562
13 December	393/1350	291/857	267/785
14 December	212/1224	126/833	111/747
15 December	0/1020	0/711	0/657
16 December	0/1121	0/633	0/569
17 December	368/406	206/323	194/298
18 December	324/683	195/317	172/279
19 December	403/0	249/0	223/0
Total	2759/6742	1719/4295	1544/3897

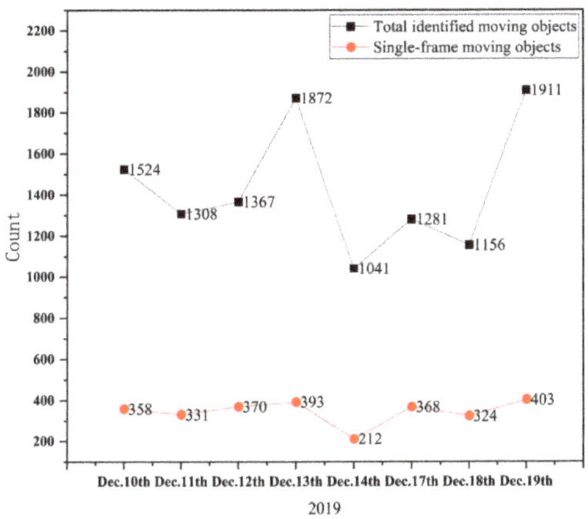

(a)

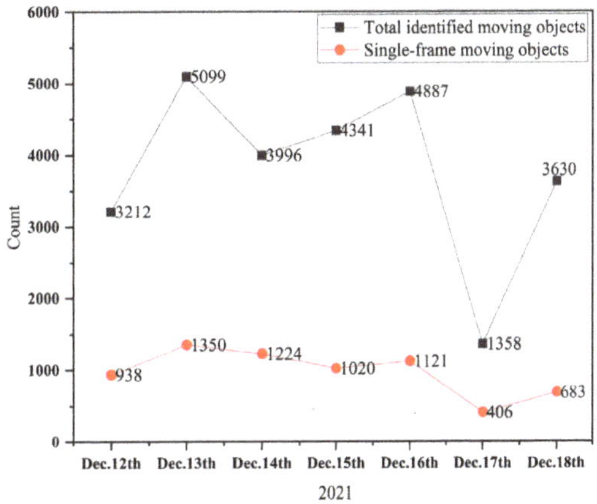

(b)

Figure 11. The results of total identified moving objects and single-frame moving objects for each date: (**a**) in 2019; (**b**) in 2021.

4. Conclusions

The GWAC system captures a significant number of meteors, and the detection of these meteors is crucial for further research on their brightness and mass information. In comparison to international meteor systems, the GWAC system exhibit characteristics such as a V-magnitude of 16, the 150 deg^2 field of view for each camera, and single-station observation. These characteristics present challenges for meteor detection in the GWAC system. Current meteor detection algorithms are limited to the mini-GWAC system data, resulting in issues including false detection, missed detection, incorrect object clustering, and mismatched object tracking within the GWAC system. We proposed a meteor detection algorithm applicable to the GWAC system based on the algorithm developed for the mini-GWAC system. Experimental results demonstrate that the proposed algorithm achieves higher accuracy and faster detection speed, with the accuracy of meteor detection reaching up to 90.27%. For future research, we could entail addressing image noise resulting from factors such as camera shake in the GWAC system to further improve the accuracy.

Author Contributions: Conceptualization, Y.C. and G.L.; methodology, Y.C. and G.L.; software, Y.C.; validation, Y.C.; formal analysis, Y.C. and G.L.; investigation, Y.C.; resources, G.L.; data curation, G.L.; writing—original draft preparation, Y.C., Q.S. and M.L.; writing—review and editing, G.L., C.L. and B.Q.; visualization, B.Q.; supervision, C.L. and B.Q.; project administration, B.Q.; funding acquisition, B.Q. All authors have read and agreed to the published version of the manuscript.

Funding: This research was funded by the Natural Science Foundation of Tianjin, grant number 22JCYBJC00410, and the Joint Research Fund in Astronomy, National Natural Science Foundation of China, grant number U1931134.

Data Availability Statement: The data presented in this study are available on request from the corresponding author. The data are not publicly available due to privacy restrictions.

Conflicts of Interest: The authors declare no conflict of interest.

References

1. Brownlee, D.; Joswiak, D.; Matrajt, G. Overview of the rocky component of Wild 2 comet samples: Insight into the early solar system, relationship with meteoritic materials and the differences between comets and asteroids. *Meteorit. Planet. Sci.* **2012**, *47*, 453–470. [CrossRef]
2. Hu, Z. Some progress in meteor physics. *Prog. Astron.* **1996**, *14*, 19–28.
3. Nascimento, J.C.; Abrantes, A.J.; Marques, J.S. An algorithm for centroid-based tracking of moving objects. In Proceedings of the 1999 IEEE International Conference on Acoustics, Speech, and Signal Processing, ICASSP99 (Cat. No.99CH36258), Phoenix, AZ, USA, 15–19 March 1999; Volume 3306, pp. 3305–3308.
4. Gural, P.S. Advances in the meteor image processing chain using fast algorithms, deep learning, and empirical fitting. *Planet. Space Sci.* **2020**, *182*, 104847. [CrossRef]
5. Gural, P. A fast meteor detection algorithm. In Proceedings of the International Meteor Conference, Egmond, The Netherlands, 2–5 June 2016; p. 96.
6. Gural, P.S. *MeteorScan Documentation and User's Guide*; Sterling VA, USA, 1999.
7. Molau, S. The Meteor Detection Software METREC. In Proceedings of the International Meteor Conference, 17th IMC, Stara Lesna, Slovakia, 20–23 August 1998; pp. 9–16.
8. Fujiwara, Y.; Nakamura, T.; Ejiri, M.; Suzuki, H. An automatic video meteor observation using UFO Capture at the Showa Station. In Proceedings of the Asteroids, Comets, Meteors 2012, Niigata, Japan, 16–20 May 2012; p. 6096.
9. Brown, P.; Weryk, R.; Kohut, S.; Edwards, W.; Krzeminski, Z. Development of an All-Sky video meteor network in southern ontario, Canada the ASGARD system. *WGN J. Int. Meteor Organ.* **2010**, *38*, 25–30.
10. Blaauw, R.; Cruse, K.S. Comparison of ASGARD and UFOCapture. In Proceedings of the International Meteor Conference, 30th IMC, Sibiu, Romania, 15–18 September 2011; pp. 44–46.
11. Molau, S. The AKM video meteor network. In Proceedings of the Meteoroids 2001 Conference, Kiruna, Sweden, 6–10 August 2001; pp. 315–318.
12. Jenniskens, P.; Gural, P.S.; Dynneson, L.; Grigsby, B.J.; Newman, K.E.; Borden, M.; Koop, M.; Holman, D. CAMS: Cameras for All sky Meteor Surveillance to establish minor meteor showers. *Icarus* **2011**, *216*, 40–61. [CrossRef]
13. Kornoš, L.; Koukal, J.; Piffl, R.; Tóth, J. EDMOND Meteor Database. In Proceedings of the International Meteor Conference, Poznan, Poland, 22–25 August 2013; pp. 23–25.
14. Karpov, S.; Orekhova, N.; Beskin, G.; Biryukov, A.; Bondar, S.; Ivanov, E.; Katkova, E.; Perkov, A.; Sasyuk, V. Meteor observations with Mini-Mega-TORTORA wide-field monitoring system. *arXiv* **2016**, arXiv:1602.07977.
15. Weryk, R.J.; Campbell-Brown, M.D.; Wiegert, P.A.; Brown, P.G.; Krzeminski, Z.; Musci, R. The Canadian Automated Meteor Observatory (CAMO): System overview. *Icarus* **2013**, *225*, 614–622. [CrossRef]
16. Feng, T.; Du, Z.; Sun, Y.; Wei, J.; Bi, J.; Liu, J. Real-time anomaly detection of short-time-scale GWAC survey light curves. In Proceedings of the 2017 IEEE International Congress on Big Data (BigData Congress), Honolulu, HI, USA, 25–30 June 2017; pp. 224–231.
17. Wan, M.; Wu, C.; Zhang, Y.; Xu, Y.; Wei, J. A pre-research on GWAC massive catalog data storage and processing system. *Astron. Res. Technol.* **2016**, *20*, 373–381.
18. Yang, C.; Weng, Z.; Meng, X.; Ren, W.; Xin, R.; Wang, C.; Du, Z.; Wan, M.; Wei, J. Data management challenges and real-time processing technologies in astronomy. *J. Comput. Res. Dev.* **2017**, *54*, 248–257. [CrossRef]
19. Götz, D.; Collaboration, T. SVOM: A new mission for Gamma-Ray Bursts studies. *Mem. Della Soc. Astron. Ital. Suppl.* **2012**, *21*, 162.
20. Bi, J.; Feng, T.; Yuan, H. Real-time and short-term anomaly detection for GWAC light curves. *Comput. Ind.* **2018**, *97*, 76–84. [CrossRef]
21. Gural, P.S. Algorithms and software for meteor detection. In *Advances in Meteoroid and Meteor Science*; Trigo-Rodríguez, J.M., Rietmeijer, F.J.M., Llorca, J., Janches, D., Eds.; Springer: New York, NY, USA, 2008; pp. 269–275.

22. Xu, Y.; Wang, J.; Huang, M.; Wei, J. An algorithm of selection of meteor candidates in GWAC system. *Astron. Res. Technol.* **2019**, *16*, 478–487. [CrossRef]
23. Yang, Q.; Feng, X.; Li, Z.; Sun, X.; Lin, X.; Wang, J. Dynamic object detection algorithm based on improved inter frame difference. *J. Heilongjiang Univ. Sci. Technol.* **2022**, *32*, 779–783.
24. Romanengo, C.; Falcidieno, B.; Biasotti, S. Hough transform based recognition of space curves. *J. Comput. Appl. Math.* **2022**, *415*, 114504. [CrossRef]
25. Fernandes, L.A.F.; Oliveira, M.M. Real-time line detection through an improved Hough transform voting scheme. *Pattern Recognit.* **2008**, *41*, 299–314. [CrossRef]
26. Zheng, F.; Luo, S.; Song, K.; Yan, C.; Wang, M. Improved lane line detection algorithm based on Hough transform. *Pattern Recognit. Image Anal.* **2018**, *28*, 254–260. [CrossRef]
27. Bednář, J.; Krauz, L.; Páta, P.; Koten, P. Meteor cluster event indication in variable-length astronomical video sequences. *Mon. Not. R. Astron. Soc.* **2023**, *523*, 2710–2720. [CrossRef]
28. Yang, Y.; Yin, C.; Liu, B.; Wang, D.; Zhang, D.; Liang, Q. Automatic recognition method research of pointer meter based on Hough transform and feature clustering. *Mach. Des. Res.* **2019**, *35*, 7–11. [CrossRef]
29. Zhu, S.; Gao, J.; Li, Z. Video object tracking based on improved gradient vector flow snake and intra-frame centroids tracking method. *Comput. Electr. Eng.* **2014**, *40*, 174–185. [CrossRef]
30. Subasinghe, D.; Campbell-Brown, M.D.; Stokan, E. Physical characteristics of faint meteors by light curve and high-resolution observations, and the implications for parent bodies. *Mon. Not. R. Astron. Soc.* **2016**, *457*, 1289–1298. [CrossRef]

Disclaimer/Publisher's Note: The statements, opinions and data contained in all publications are solely those of the individual author(s) and contributor(s) and not of MDPI and/or the editor(s). MDPI and/or the editor(s) disclaim responsibility for any injury to people or property resulting from any ideas, methods, instructions or products referred to in the content.

universe

MDPI

Article

Molecular-Clump Detection Based on an Improved YOLOv5 Joint Density Peak Clustering

Jin-Bo Hu [1,2], Yao Huang [1,3,*], Sheng Zheng [1,3,*], Zhi-Wei Chen [4], Xiang-Yun Zeng [1,3] and Xiao-Yu Luo [1,3] and Chen Long [1,3]

1 Center for Astronomy and Space Sciences, China Three Gorges University, Yichang 443000, China; jinbohu@ctgu.edu.cn (J.-B.H.); xyzeng2018@ctgu.edu.cn (X.-Y.Z.); lxy@ctgu.edu.cn (X.-Y.L.); longc25@ctgu.edu.cn (C.L.)
2 College of Computer and Information Technology, China Three Gorges University, Yichang 443000, China
3 College of Science, China Three Gorges University, Yichang 443000, China
4 Purple Mountain Observatory, Chinese Academy of Sciences, Nanjing 210023, China; zwchen@pmo.ac.cn
* Correspondence: huangyao@ctgu.edu.cn (Y.H.); zsh@ctgu.edu.cn (S.Z.)

Abstract: The detection and analysis of molecular clumps can lead to a better understanding of star formation in the Milky Way. Herein, we present a molecular-clump-detection method based on improved YOLOv5 joint Density Peak Clustering (DPC). The method employs a two-dimensional (2D) detection and three-dimensional (3D) stitching strategy to accomplish the molecular-clump detection. In the first stage, an improved YOLOv5 is used to detect the positions of molecular clumps on the Galactic plane, obtaining their spatial information. In the second stage, the DPC algorithm is used to combine the detection results in the velocity direction. In the end, the clump candidates are positioned in the 3D position-position-velocity (PPV) space. Experiments show that the method can achieve a high recall of 98.41% in simulated data made up of Gaussian clumps added to observational data. The efficiency of the strategy has also been demonstrated in experiments utilizing observational data from the Milky Way Imaging Scroll Painting (MWISP) project.

Keywords: YOLOv5; density peak clustering; molecular clouds; clump detection

1. Introduction

As one of the most essential components of the interstellar medium, molecular clouds consist primarily of molecular gas mixed with small amounts of atoms, ions, and dust [1]. The scale of molecular clouds spans from giant clouds with tens of pc to dense clumps with less than 0.1 pc [2]. Molecular clouds usually present complex and hierarchical structures. Clumpiness is a universal property of molecular clouds and plays a key role in understanding the fragmentation of cloud complexes into sub-clouds finally small enough to form individual stars [3]. Modern astronomy recognizes dense clumps in the molecular clouds as the key sites of star formation [4]. Thus, the detection and analysis of dense clumps in molecular clouds can better elucidate star formation and matter cycling in the Milky Way [5].

Several researchers have already studied automatic detection algorithms for molecular clumps. In 1990, Stutzki and Guesten (1990) [3] proposed the GaussClumps algorithm and applied it to the M17 molecular cloud to detect 170 clumps. In 1994, Williams et al. (1994) [6] proposed the ClumpFind algorithm and tested it on RMC and MMC. They detected 83 clumps in the RMC and 78 clumps in the two observed regions of MMC. In 2015, Berry (2015) [7] proposed the FellWalker algorithm, which has been utilized by Kirk et al. (2016) [8] to detect a total of 915 clumps in three subregions of the Orion B molecular cloud and obtain basic physical information, such as the fluxes and sizes of these clumps. In 2022, Luo et al. (2022) [9] proposed the LDC algorithm, which has detected 658 clumps in the M16 molecular cloud. Astronomers need to set the initial parameters of these detection algorithms according to the observational instrument and the morphological features of the target during detection. The detection results are closely related to the parameter settings. Before detection, astronomers need to repeatedly adjust the parameters to optimize the algorithm or manually verify the detection results and thus ensure accuracy. To accomplish the task of large-scale molecular clump census from the rapidly growing CO molecular-cloud observation, the research, and development of detection algorithms with higher efficiency are needed.

With the rapid development of artificial intelligence technology, deep-learning methods represented by convolutional neural networks (CNNs) have extensively been used in image and machine vision [10–15]. Deep learning also provides a viable method for extracting essential features of astronomical data. Kim and Brunner (2017) [16] proposed a CNN to classify stars and galaxies. Gonz´alez et al. (2018) [17] implemented the task of galaxy detection and classification by using the YOLO target-detection algorithm. Leung and Bovy (2019) [18] designed a Bayesian neural network to measure multi-element abundances in stellar spectra. Xie et al. (2021) [19] applied a convolutional neural model to stellar spectra for detecting stars with low metal content. He et al. (2021) [20] proposed a target-detection network based on YOLOv4 to detect sources in SDSS images and a classification model called APSCnet to classify sources. Yi et al. (2022) [21] proposed an automatic detection model based on a deep-learning approach for automatically detecting low-surface-brightness galaxies from SDSS images. Cao et al. (2023) [22] used an improved Faster R-CNN framework based on deep learning to detect L-dwarf from SDSS images automatically. These new techniques enable the automatic analysis of astronomical data and demonstrate the feasibility of supervised deep learning for astronomical data mining. Therefore, an automatic clump-detection algorithm that extracts the desired types of targets directly from the observational data through supervised feature learning from clump morphological properties can be similarly designed for CO survey data.

The Milky Way Imaging Scroll Painting (MWISP) project [23] is a large-scale CO survey project carried out by the Purple Mountain Observatory of the Chinese Academy of Sciences by using the DLH-13.7 m millimeter-wave radio telescope, which simultaneously observes ^{12}CO, ^{13}CO, and $C^{18}O$(1-0) lines emission. The first phase of MWISP has acquired a large number of molecular-cloud observational data, providing sufficient training samples for the deep-learning-based clump detection algorithm. This paper proposes a clump-detection method based on an improved YOLOv5 joint Density Peak Clustering (DPC) [24]. The method initially locates the clumps in the position-position (PP) coordinate representing the Galactic longitude and latitude by using the improved YOLOv5. Subsequently, DPC is used to cluster the clumps in the velocity direction, thereby ultimately realizing the clump detection in position-position-velocity (PPV) three-dimensional (3D) space. This method utilizes supervised deep learning to detect clumps by labeling the areas of interest. The detection results are directly related to the labeling scheme and feature learning of the data. During detection, the 3D information where the target candidate is located can be obtained quickly with far fewer parameters tuning during detection.

The paper is organized as follows. Section 2 describes the generation of the experimental data. Section 3 details the theory of the clump-detection algorithm based on the improved YOLOv5 joint DPC. Section 4 presents the training process and the main experimental results in PPV space. Finally, conclusions are drawn in Section 5.

2. Data

To better train the improved YOLOv5 and comprehensively evaluate the performance of the proposed method, a large number of clumps are required. Synthesized data are designed for the quantitative evaluation of detection performance. Synthesized data are composed by randomly adding simulated clumps to the observational data background. The simulated clumps are generated by the 3D Gaussian model, and the background is the observational ^{13}CO(1-0) line emission data obtained by MWISP. The simulated clumps, the observational data, and the synthesized data are described separately below.

2.1. Simulated Clumps

Localized high densities characterize the clumps, and Gaussian functions can be used to describe their intensity distributions [3]. Simulated clumps can be generated utilizing the 3D Gaussian model with several parameters [25], as shown in Table 1. The 3D Gaussian model is described as:

$$f(x,y,v) = A\exp\left\{-\left[(\frac{\cos^2\theta}{2\sigma_x^2}+\frac{\sin^2\theta}{2\sigma_y^2})(x-x_0)^2 +(\frac{\cos^2\theta}{2\sigma_y^2}-\frac{\sin^2\theta}{2\sigma_x^2})(x-x_0)(y-y_0) +(\frac{\sin^2\theta}{2\sigma_y^2}+\frac{\cos^2\theta}{2\sigma_y^2})(y-y_0)^2+\frac{(v-v_0)^2}{2\sigma_v^2}\right]\right\} \quad (1)$$

where A represents the peak intensity of the clump, σ_x, σ_y, σ_v represent the standard deviations on the Galactic longitude, the Galactic latitude, and the velocity, respectively, (x_0, y_0, v_0) represents the position of the center of mass of the clump, and θ represents the rotation angle on the Galactic plane. To enable generated simulated clumps with the parameter-distribution characteristics of the observational data, the statistics of the clump parameters detected in the ^{13}CO(1-0) line emission data of the M16 region obtained by MWISP are regarded as the input into the 3D Gaussian model. The ^{13}CO(1-0) line emission of M16 region ranges $15°15' < l < 18°15'$, $0° < b < 1°30'$. LDC [9] algorithm was applied to detect the ^{13}CO(1-0) line emission of M16 and a total of 658 clumps have been detected. We counted the morphological parameters of these clumps and obtained the parameters range in Table 1. Five thousand simulated clumps were randomly generated, and fluxes were calculated according to the parameters in Table 1. The flux distribution is shown in Figure 1. These simulated clumps maintain consistency with the MWISP observational data and provide a more realistic representation of the detection performance.

Table 1. Parameters of the 3D Gaussian model.

Parameter Name	Explanation	Range
Peak	Peak intensity of the clump	[0.7, 15]
σ_x	Standard deviation on the Galactic longitude	[1, 4] × 2.3548
σ_y	Standard deviation on the Galactic latitude	[1, 4] × 2.3548
σ_v	Standard deviation on the velocity direction	[1, 7] × 2.3548
(x_0, y_0, v_0)	Position of the center of mass of the clump	Randomization
θ	Rotation angle on the Galactic plane	0°–180°

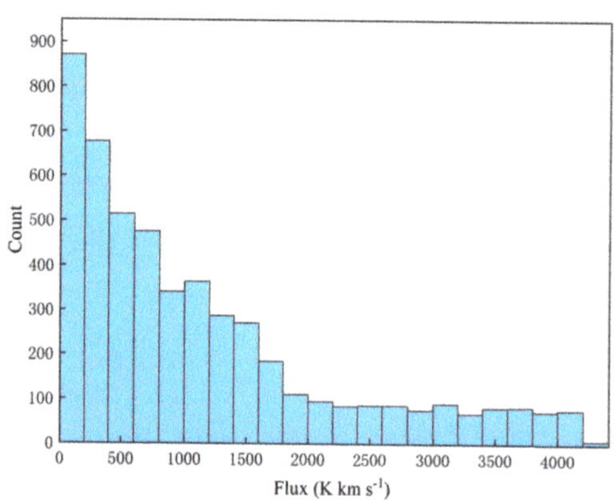

Figure 1. Distribution of the number of fluxes of 5000 randomly generated simulated clumps.

2.2. *Observational Data*

Three typical regions within the first, second, and third Galactic quadrants of the $^{13}CO(1\text{-}0)$ line emission data obtained from MWISP are selected as the background for generating the synthesized data. The typical noise level of $^{13}CO(1\text{-}0)$ line emission is about 0.23 K with a channel width of 0.167 km s^{-1}. Three regions all contain information in the three dimensions of Galactic longitude, Galactic latitude, and velocity. Due to the structure of the spiral arm of the Milky Way, different quadrants contain different spiral arms and have different gas distributions. The range of the Galactic plane is $3^\circ \times 2^\circ$ and the velocity range is 70 km s^{-1}, of all three backgrounds, corresponding with the size of $361 \times 241 \times 424$ pixels. The regions selected in the first Galactic quadrant range $13^\circ < l < 16^\circ$ and $-1^\circ30' < b < 30'$, and the velocity ranges 0 km s^{-1} $< vs. <$ 70 km s^{-1}. It belongs to the inner Milky Way area, in the direction of the galactic center. The regions selected in the second Galactic quadrant range $101^\circ < l < 104^\circ$ and $2^\circ < b < 4^\circ$, and the velocity ranges -60 km s^{-1} $< vs. <$ 10 km s^{-1}. The regions selected in the third Galactic quadrant range $184^\circ30' < l < 187^\circ30'$ and $-1^\circ < b < 1^\circ$, and the velocity ranges -10 km s^{-1} $< vs. <$ 60 km s^{-1}. They both belong to the outer Milky Way area. The density of the three regions is different. The background gas is dense in the first Galactic quadrant region and sparse in the third Galactic quadrant region. In the second Galactic quadrant region, the density of the background is between the first and the third Galactic quadrants. Different gas densities can reflect the detection performance of the detection algorithm in different signal-to-noise ratio environments. Figure 2 shows the velocity-integrated intensity maps for each selected background region.

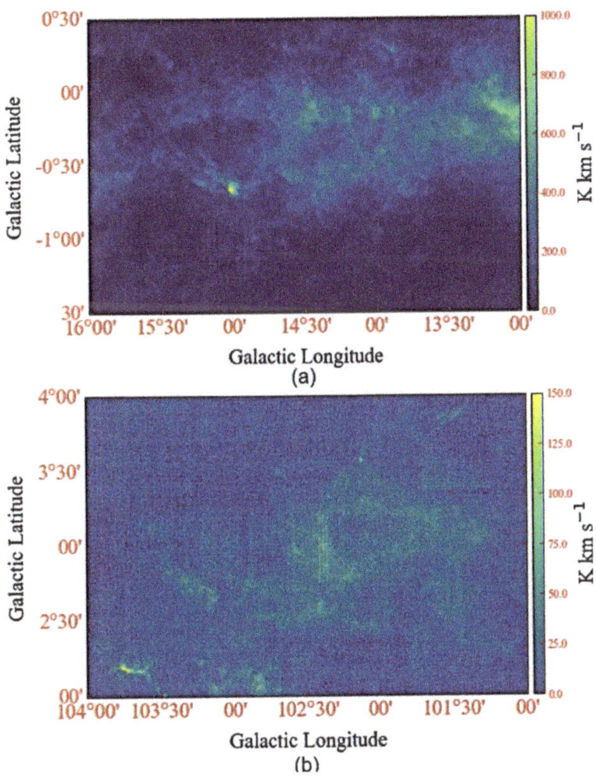

Figure 2. *Cont.*

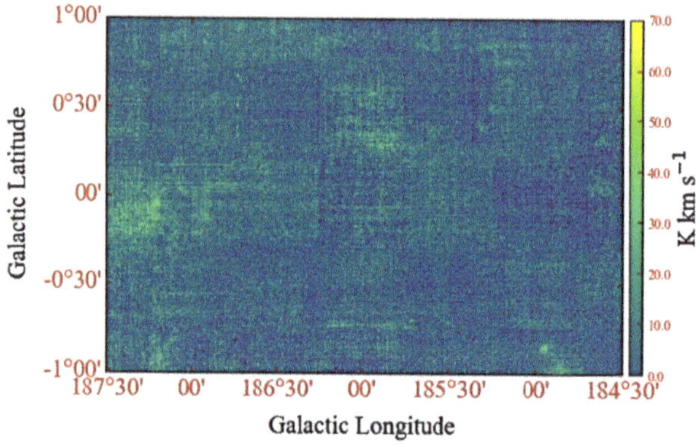

(c)

Figure 2. Velocity−integrated intensity maps for selected background regions. (**a**) the first Galactic quadrant with $13° < l < 16°$ and $-1°30' < b < 30'$; the velocity ranges 0 km s^{-1} $< vs. <$ 70 km s^{-1}. (**b**) the second Galactic quadrant with $101° < l < 104°$ and $2° < b < 4°$; the velocity ranges -60 km s^{-1} $< vs. <$ 10 km s^{-1}. (**c**) the third Galactic quadrant with $184°30' < l < 187°30'$ and $-1° < b < 1°$; the velocity ranges -10 km s^{-1} $< vs. <$ 60 km s^{-1}.

2.3. Synthesized Data

One hundred simulated clumps obtained using the 3D Gaussian model are added randomly to the designated region mentioned in Section 2.2 to generate synthesized data. The added simulated clumps are uniformly distributed throughout all regions. This process is repeated three times to create the first, second, and third Galactic quadrants' synthesized data. To ensure that each synthesized data item has 100 complete simulated clumps, it should avoid boundary regions and clump fusion during the addition process. Velocity-integrated intensity maps for three Galactic quadrants' synthesized data are shown in Figure 3. Synthesized data also contains information in the three dimensions of Galactic longitude, Galactic latitude, and velocity. The size of synthesized data is the same as its corresponding background region size. The range of the Galactic plane is $3° \times 2°$ and the velocity range is 70 km s^{-1}, of all synthesized data, corresponding with the size of $361 \times 241 \times 424$ pixels.

As the synthesized data are generated, information about each added simulated clump is recorded in a clump parameter table for each synthesized data. The information contained in the table is shown in Table 2, where ID denotes the clump number. Peak1, Peak2, and Peak3 indicate the peak intensity coordinates. Cen1, Cen2, and Cen3 indicate the center of mass coordinates. Size1, Size2, and Size3 denote the axial lengths. θ represents the rotation angle on the Galactic plane. Sum denotes the total flux. Peak indicates the peak intensity.

Once the detection is complete, the results are compared with the clump parameter tables. The algorithm performance can then be quantitatively evaluated.

Table 2. Parameter information table of simulated clumps.

Parameter Name	Explanation
ID	Clump number
Peak1, Peak2, Peak3	Peak coordinates of clumps
Cen1, Cen2, Cen3	Coordinates of the center of mass of clumps
Size1, Size2, Size3	Axis lengths of clumps in the Galactic plane, and velocity direction
θ	Rotation angles of clumps on the Galactic plane
Sum	Total flux of clumps
Peak	Peak intensity of clumps

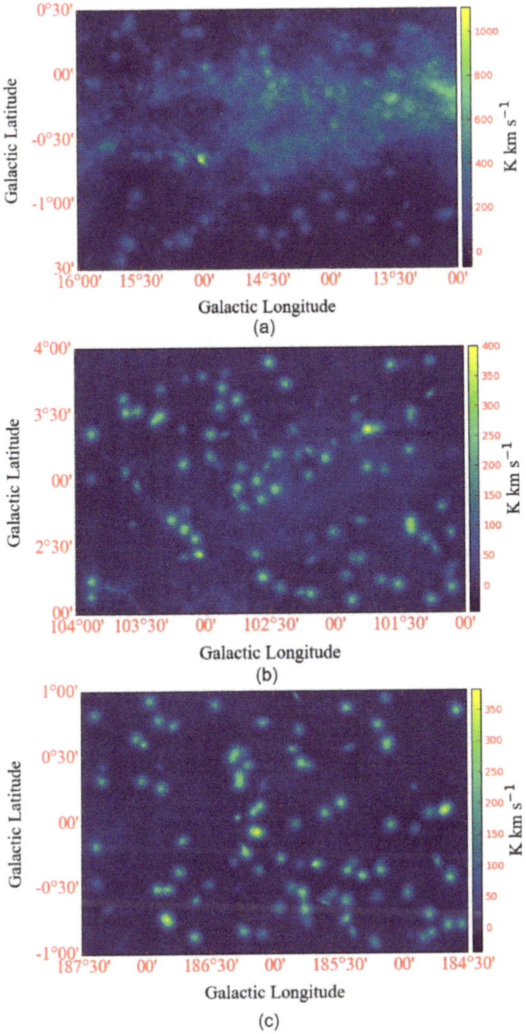

Figure 3. Velocity−integrated intensity maps of the synthesized data: (**a**) the first Galactic quadrant synthesized data. (**b**) the second Galactic quadrant synthesized data. (**c**) the third Galactic quadrant synthesized data.

3. Method

This paper presents a new method based on an improved YOLOv5 joint DPC for detecting molecular clumps in PPV data. Detection is divided into two stages. In the first stage, the improved YOLOv5 is used to detect the clumps in two-dimensional (2D) space and obtain their position on the Galactic plane. In the second stage, DPC is used to combine the detection results obtained in the first stage in the velocity direction, thereby helping achieve the final clump detection in 3D space. The code has been shared on Github [1].

3.1. The Improved YOLOv5 − Molecular Clump Detection (MCD)-YOLOv5

YOLOv5 is an extensively used target-detection tool that combines speed and accuracy [26–29]. One of its key advantages is that it is a one-stage detection algorithm. It provides faster inference and detection speeds, making it useful for handling large amounts of molecular cloud data.

YOLOv5 consists of a backbone, neck, and YOLO head. The backbone can extract the information of the input. The neck can extract features from the backbone to improve the performance of the network. The YOLO head outputs the results of the network predictions. Although YOLOv5 is a strong performer in many target-detection tasks, it has limitations in detecting small targets which are often low in pixels, small in percentage, easy to overlap, and difficult to distinguish. The molecular-cloud data have clumps with weak intensity which may be obscured by the molecular-cloud background. Owing to the rotation angle, some clumps also appear small on the Galactic plane. To enhance the ability of YOLOv5 to detect these small clumps, we introduce the Coordinate Attention (CA) module and Normalized Wasserstein Distance (NWD) loss function to YOLOv5. This improved version, called Molecular clump detection (MCD)-YOLOv5, and its architecture is shown in Figure 4.

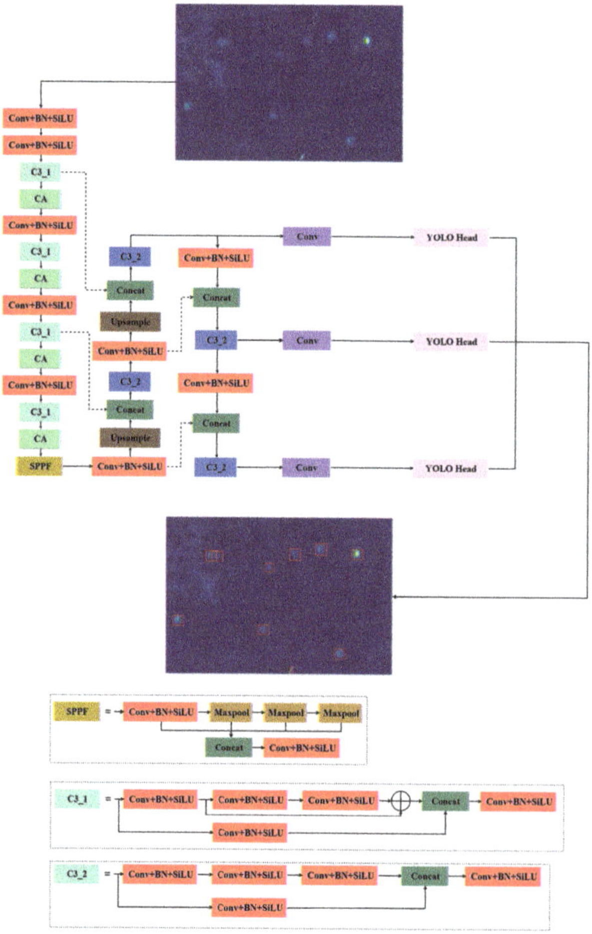

Figure 4. Architecture of MCD-YOLOv5. MCD-YOLOv5 is mainly composed of three parts: backbone, neck, and YOLO head. The backbone consists of CBS, C3_1, and CA modules. CBS is a composite convolution module, which encapsulates a convolutional layer, a batch normalization layer, and the SiLU activation function. MCD-YOLOv5 contains two types of C3 modules; C3_1 is applied in the backbone and C3_2 is applied in the neck. The C3_1 module and C3_2 module both consist of CBS.

In C3_1, the input feature map passes through three branches, while in C3_2, it only passes through two branches. The branches are finally spliced by channel and then output through a CBS module. We add a CA module after each layer of the C3_1. The neck consists of SPPF(Spatial Pyramid Pooling-Fast), and CSP-PAN(CSP-Path Aggregation Network) modules. SPPF is a spatial pyramid pooling module. SPPF encapsulates a CBS module and three maximum pooling layers. Three pooling results with input feature maps are spliced by channel and passed through a CBS module. CSP-PAN is composed of CBS and C3_2, and the feature fusion of different feature layers is realized through upsampling and downsampling, which solves the target multi-scale problem to a certain extent. The main part of the YOLO head is three detectors, that is, using mesh-based anchors to detect objects on feature maps at different scales.

3.1.1. Coordinate Attention

To improve YOLOv5's ability to recognize the clumps with weak intensity in slices, a CA module [30] is utilized. The architecture of the CA module is shown in Figure 5. CA is a lightweight attention module, which encodes each channel of the feature map along the X and Y directions. The resulting feature maps are combined and transformed via convolutional transform to create intermediate feature maps. These feature maps are then divided into separate tensors, which are again transformed and expanded to become the value of the attention weight assignment. The CA module is added after each layer of the C3_1 module in YOLOv5, thereby increasing the number of layers from 10 to 14. This action enables a more accurate identification of regions of interest during the feature extraction of clumps.

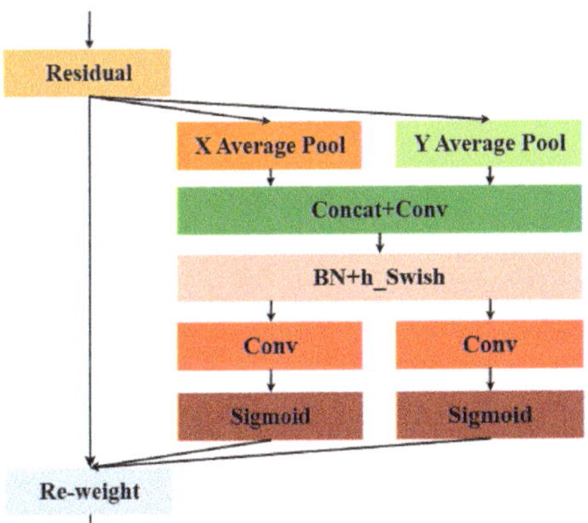

Figure 5. Architecture of Coordinate Attention module. The module is made of two average pooling layers, a convolution layer with concat operation, a batch normalization layer with an h_swish activation function, two convolution layers, and two sigmoid activation functions. The average pooling layers encode each channel of the feature map along the X and Y directions. The resulting feature maps are combined and transformed via the convolution layer with concat operation and the batch normalization layer with an h_swish activation function to create intermediate feature maps. These feature maps are then divided into separate tensors, which are again transformed and expanded by the convolution layer and sigmoid activation function to become the value of the attention weight assignment.

3.1.2. Normalized Wasserstein Distance

Detecting small clumps is challenging owing to limited appearance information, conferring difficulty in recognizing features. The use of Intersection over Union (IOU) [31] as a metric in YOLOv5 to measure the goodness of the generated anchor frames during detection also results in the poor detection of small targets. Wang et al. (2022) [32] analyzed the sensitivity of IoU to location deviations of tiny objects, and proposed NWD as a better metric for measuring the similarity between two bounding boxes. To improve the detection performance of YOLOv5 for small target clumps, the NWD loss function is introduced. The expressions for NWD and the loss function based on NWD are presented as follows:

$$W_2^2(N_p, N_g) = \left\| \left(\left[cx_p, cy_p, \frac{w_p}{2}, \frac{h_p}{2} \right]^T, \left[cx_g, cy_g, \frac{w_g}{2}, \frac{h_g}{2} \right]^T \right) \right\|_2^2 \tag{2}$$

$$NWD(N_p, N_g) = \exp\left(-\frac{\sqrt{W_2^2(N_p, N_g)}}{C} \right) \tag{3}$$

$$L_{NWD} = 1 - NWD(N_p, N_g) \tag{4}$$

where N_p is the Gaussian distribution model of the prediction frame P; N_g is the Gaussian distribution model of the gt frame G; (cx_p, cy_p, w_p, h_p) is the coordinates of the upper left corner of the prediction frame P and the width and height information; (cx_g, cy_g, w_g, h_g) represents the coordinates of the upper left corner of the gt frame G and the width and size information, respectively; and $W_2^2(N_p, N_g)$ is the second-order Wasserstein distance between N_p and N_g. C is a constant. This paper uses the original Complete IoU Loss (CIOU) [33] of YOLOv5 and NWD loss in the molecular-clump-detection task. The weight ratio of CIOU loss and NWD loss is 7:3.

3.2. Density Peak Clustering Algorithm

The DPC algorithm was proposed by Rodriguez & Laio (2014) [24] in 2014. DPC assumes that points with low local density surround the cluster centers, which are relatively distant from others. This feature is very similar to clumps embedded in the molecular cloud with lower average density and localized density enhancement. Accordingly, this paper uses DPC to complete the splicing of clumps in the velocity direction.

DPC is implemented by dividing the data points within a specific range into a region followed by calculating the density ρ_i and distance δ_i of each data point within that region:

$$\delta_i = \begin{cases} \min\limits_{j:\rho_j \geq \rho_i} d_{ij}, & if\ \rho_j \geq \rho_i \\ \max\limits_{j=1,2,\ldots,n} d_{ij}, & otherwise \end{cases} \tag{5}$$

d_{ij} denotes the Euclidean distance between data points i and j. When a point with a higher density than the current data point exists, the distance of this point is the minimum of all data points denser than the point. After setting the thresholds for density and distance, clustering centers can be determined from these data points. Clustering can also be accomplished by assigning points not clustering centers to the nearest clustering centers based on the distances of other data points from the clustering centers.

After MCD-YOLOv5, the clumps are extracted in the slices by the coordinate position information obtained from the detection, and the center of mass of the extracted region is calculated by the adaptive threshold segmentation [34]. Through the position coordinates (x_i, y_i) of the center of mass and its velocity channel information v_i, the intensity I_i of the corresponding position in the synthesized data is also extracted. These pieces of information are used in this paper to represent the density of each input data point with the following expression:

$$\rho_i = I_i \tag{6}$$

The inputs and outputs of DPC are listed in Table 3. DPC clusters the centers with position and density output by MCD-YOLOv5 and divides them into different categories. The final number of categories is the number of clumps detected.

Table 3. Input and output data for DPC.

Description	Parameters Name	Explanation
Input	x_i	Galactic longitude coordinates of the center of mass in the region detected by MCD-YOLOv5
	y_i	Galactic latitude coordinates of the center of mass in the region detected by MCD-YOLOv5
	v_i	Channels in the velocity direction detected by MCD-YOLOv5
	I_i	Intensity of (x_i, y_i, v_i) in the synthesized data
Output	numClust	Number of clustered clumps categories

3.3. MCD-YOLOv5 Joint DPC

The flowchart of the clump-detection method based on MCD-YOLOv5 joint DPC is shown in Figure 6. The molecular cloud data are sliced in the velocity direction, resulting in 2D data along the Galactic plane. These data are inputted into MCD-YOLOv5 for detection. Once all slices have been detected, MCD-YOLOv5 outputs the coordinate position information of the detected clumps. The center of mass of each detected clump can be calculated based on MCD-YOLOv5 output information. Using the center of mass and the velocity channel size where the corresponding slice is located, the intensity value of the related part on the molecular cloud data is obtained as the input to DPC. DPC clusters the centers of mass and classifies them into different categories. The region comprising clumps corresponding with center of mass points in the same category is classified as a clump. The spatial coordinates of the slices of molecular clouds belonging to the same clump in the direction of the Galactic plane and the span in the direction of velocity are statistically calculated to transform the PP data into PPV data. The output of MCD-YOLOv5 joint DPC is 3D cubic regions containing clumps. By obtaining information about each 3D region, we can calculate the centroid (Cen_i) and size ($Size_i$) of the molecular cloud clumps on different axes i $(i = 1, 2, 3)$. The definition of the centroid is as follows:

$$Cen_i = \frac{\sum_{j=1}^{n_k} (I_j \cdot x_j)}{\sum_{j=1}^{n_k} I_j} \tag{7}$$

the $Size_i$ of the clump on axis i is defined as:

$$Size_i = \sqrt{\frac{\sum_{j=1}^{n_k} (I_j \cdot x_j^2)}{\sum_{j-1}^{n_k} I_j} - \left(\frac{\sum_{j=1}^{n_k} (I_j \cdot x_j}{\sum_{j=1}^{n_k} I_j} \right)^2} \tag{8}$$

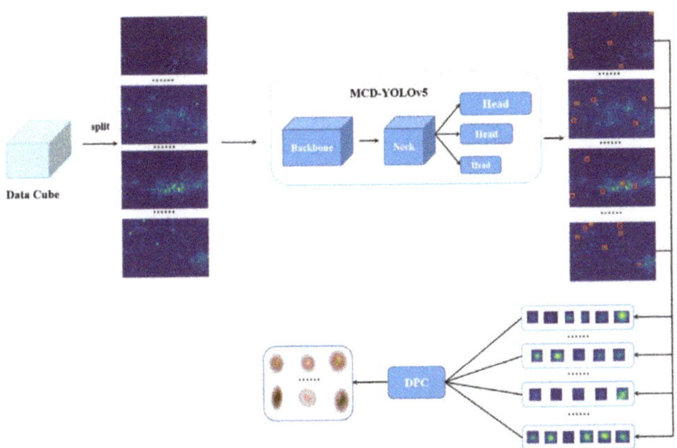

Figure 6. Flowchart of the clump-detection method based on MCD-YOLOv5 joint DPC. The four branches represent any four slices in a given data cube. Each slice achieves 2D detection by MCD-YOLOv5 to obtain the anchor information of the target on the corresponding slice. The rest slices are indicated by ellipses. Heads in different size fonts indicate different scales of detection heads.

4. Experiments and Discussion

To verify the effectiveness and reliability of the detection method based on MCD-YOLOv5 joint DPC, MCD-YOLOv5 is trained by using synthesized data from the first and third Galactic quadrants. Its performance in detecting clumps is tested using the synthesized data from the second Galactic quadrant. The training process of MCD-YOLOv5, the effectiveness of DPC, and the test results are presented in the following sections.

4.1. Evaluation of Indicators

The performance of MCD-YOLOv5 for detecting clumps on the 2D Galactic plane is quantified by using precision, recall, and average precision (AP), which are the most commonly used in target-detection tasks. Using the recall rate to evaluate the 3D detection performance of the clump-detection method based on MCD-YOLOv5 joint DPC. The precision, recall, and AP expressions are as follows:

$$precision = \frac{TP}{TP + FP}, \tag{9}$$

$$recall = \frac{TP}{TP + FN}, \tag{10}$$

$$AP = \int_0^1 P(r) \tag{11}$$

In calculating precision and recall, true positive (TP) denotes the number of clumps predicted to be clumps that are actually clumps, false positive (FP) denotes the number of clumps predicted to be clumps but are not actually clumps, and false negative (FN) denotes the number of clumps predicted to be not clumps that are actually clumps. In calculating AP, P(r) denotes the maximum precision value when the recall takes the corresponding value.

4.2. MCD-YOLOv5 Training and DPC

4.2.1. MCD-YOLOv5 Dataset Generation

The slices generated by intercepting along the velocity channel of the synthesized data are used as samples in the dataset. Each sample has only one velocity channel and contains information in two dimensions, Galactic longitude, and Galactic latitude, with a pixel size of 361 × 241. A total of 45 first-Galactic-quadrant and 45 third-Galactic-quadrant synthesized data cubes with 9000 simulated clumps to generate the dataset used for the training and validation of MCD-YOLOv5. There are a total of 37,350 samples in this dataset. The samples obtained from the first-Galactic-quadrant and third-Galactic-quadrant synthesized data are shown in Figure 7a and Figure 7b, respectively.

Considering that MCD-YOLOV5 is a supervised deep-learning method, the location information on the clumps in each sample needs to be provided. The position information of clumps in the training samples can be obtained by the clump parameter table mentioned in Section 2.3. The specific process for obtaining the label information is as follows. The center of mass (Cen1 and Cen2) of the simulated clumps on the Galactic plane is taken as the starting point. To form a rectangle that can cover the clumps, the pixel distance with Size1 is expanded upward and downward, and the pixel distance with Size2 is expanded to the left and the right. Recording the coordinate position information of the rectangles gives information about the labeling of clumps on a sample. We can obtain the position information of the same simulated clump appearing in the sample by taking the center of mass coordinate Cen3 of the simulated clump in the velocity direction as the midpoint and expanding the width of Size3 forward and backward. Figure 7 shows that the simulated clumps are accurately labeled by the above method. Figure 7c shows the labeling of the synthesized data slices in the third Galactic quadrant. Figure 7d shows the labeling of the synthesized data slices in the first Galactic quadrant.

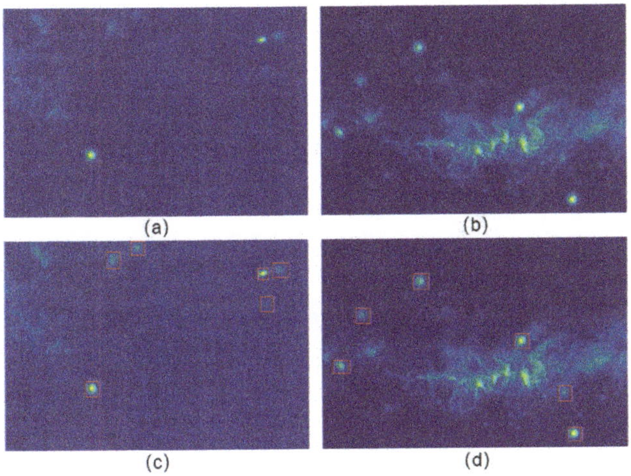

Figure 7. Examples from the MCD-YOLOv5 dataset and the annotation information: (**a**) synthesized data in the third Galactic quadrant, (**b**) synthesized data in the first Galactic quadrant, (**c**) annotation information on (**a**), and (**d**) annotation information on (**b**). The red-labeled boxes represent the clumps.

4.2.2. Training MCD-YOLOv5

To train MCD-YOLOv5 effectively, the dataset is divided into a training set and a test set at a ratio of 8:2, and takes 10% from the training set as the validation set. The training set is used to train MCD-YOLOv5, the validation set is used to verify the performance metrics of MCD-YOLOv5 during training, and the test set is used to test the detection effect of MCD-YOLOv5 after training is completed. During training, the number of training rounds is set to 300, the amount of batch training data is 128, the training momentum is 0.9, the initial learning rate is 0.001, and the weight decay is 0.0005. The stochastic gradient descent (SGD) serves the optimization function to train MCD-YOLOv5. The default parameters of YOLOV5 were used for training in this paper.

To verify the improved performance of MCD-YOLOv5 compared with YOLOv5 in detecting clumps, we train MCD-YOLOv5 and YOLOv5 on the same dataset. Figure 8 shows the variation in localization loss and confidence loss. The curve variation shows that MCD-YOLOv5 and YOLOv5 are not overfitted or underfitted, and MCD-YOLOv5 decreases more smoothly during training. The variation curves of the precision and recall on the validation set with the number of training rounds are shown in Figure 9a,b. Figure 9c demonstrates the variation of AP in MCD-YOLOv5 and YOLOv5 during the training when IOU is set to 0.5. IOU denotes the intersection and concurrency ratio of the area of the predicted frame to the area of the actual frame. In general, the predicted frames produced by the model are recognized to be correct only if the IOU is greater than or equal to the set threshold. The curve changes show that the precision of MCD-YOLOv5 does not considerably differ from that of YOLOv5, but its recall rate exceeds that of YOLOv5.

We select a slice containing smaller clumps and input it into the trained MCD-YOLOv5 and YOLOv5 at the same time for detection. Results are shown in Figure 10. By comparison, MCD-YOLOv5 detects all seven clumps in the slices, whereas YOLOv5 misses a smaller clump labeled 2. This experiment illustrates the improved small target detection performance of the MCD-YOLOv5. Table 4 shows the precision, recall, and AP of MCD-YOLOv5 and YOLOv5 in the test set. By comparison, MCD-YOLOv5 performs better than YOLOv5 in detecting clumps when processing the same molecular cloud data slices.

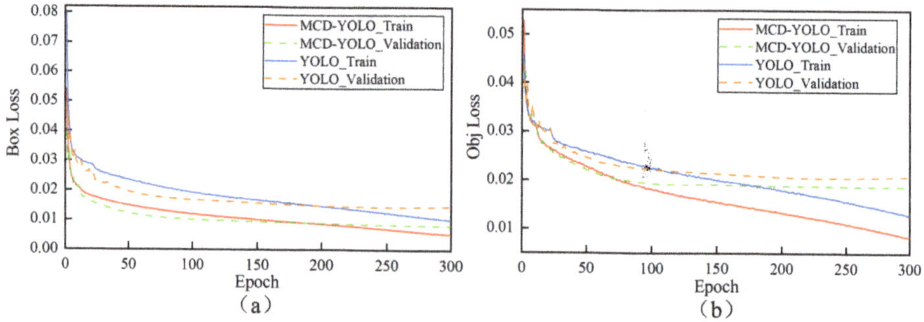

Figure 8. Comparing loss-change curves MCD-YOLOv5 and YOLOv5 training. The clump-detection task has only one category of clump, so the loss contains only localization loss and confidence loss: (**a**) variation in localization loss (Box Loss) with number of training rounds (epoch), (**b**) variation in confidence loss (Obj Loss) with number of training rounds (epoch).

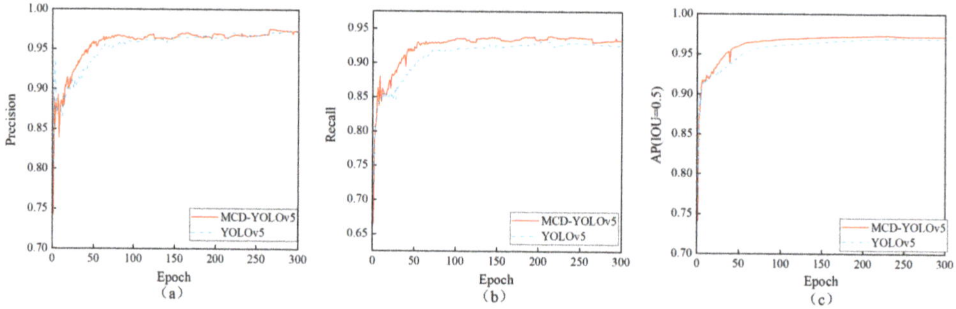

Figure 9. Variation curves of precision, recall, and AP as a function of the number of training rounds (epoch) for MCD-YOLOv5 and YOLOv5 on the validation set: (**a**) precision, and (**b**) recall, and (**c**) AP.

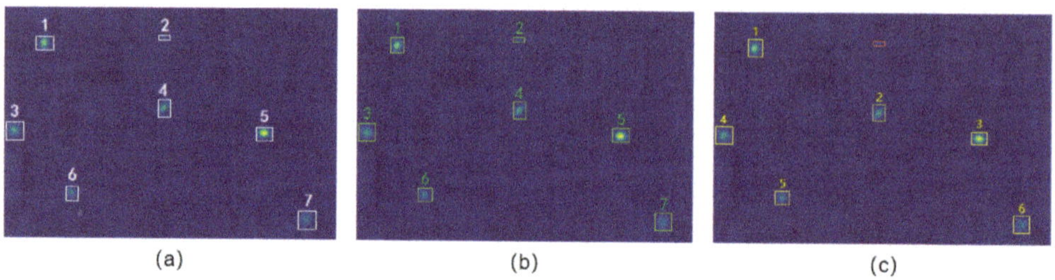

Figure 10. A typical example of detection results of MCD-YOLOv5 and YOLOv5. (**a**) Seven simulated clumps in the area. (**b**) MCD-YOLOv5 detected 7 clumps. (**c**) YOLOv5 detected 6 clumps. The red rectangular box in (**c**) represents the missed clump.

Table 4. Detection results of MCD-YOLOv5 and YOLOv5 on the test set.

Model	Precision	Recall	AP
MCD-YOLOv5	0.969	0.935	0.972
YOLOv5	0.969	0.910	0.956

4.2.3. Result of DPC

When MCD-YOLOv5 completes the detection, the center of mass of the detected clumps and their corresponding intensity information is inputted into DPC according to the method mentioned in Section 3.2. DPC then classifies the centers of mass belonging to the same clumps into the same class. The results of DPC are shown in Figure 11. The differently colored dots represent the center of mass of a clump detected by MCD-YOLOv5, the red rhombus represents a clustering center, and the green square represents the location of the center of mass of a simulated clump. If the dots representing the detected clumps have the same color, these centers of mass are classified as one clump. Figure 11 shows that most of the cluster centers obtained through DPC can correspond with the centers of mass of the simulated clumps. This finding indicates that using intensity-based DPC can classify slices belonging to the same clumps into a single class in the direction of velocity, thereby ultimately realizing the detection of 3D clumps in PPV.

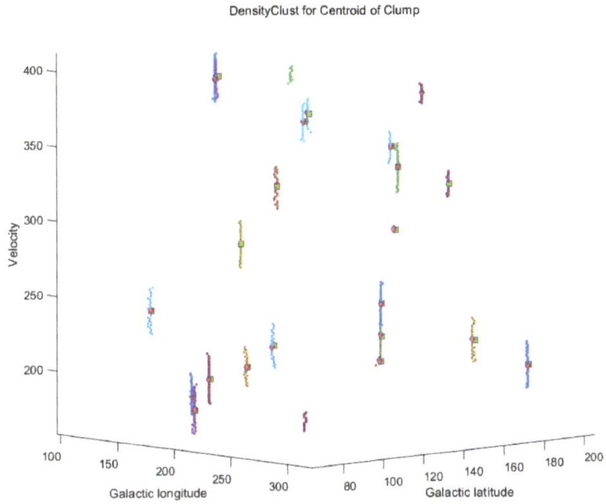

Figure 11. Display of the DPC results. The differently colored dots represent the center of mass of a clump detected by MCD-YOLOv5, the red rhombus represents a clustering center, and the green square represents the location of the center of mass of a simulated clump.

4.3. Detection Results of Second-Quadrant Synthesized Data

To test the detection performance of the proposed method based on MCD-YOLOv5 joint DPC, 100 second-Galactic-quadrant synthesized data (containing 10,000 simulated clumps) are selected for testing. To effectively evaluate the performance, we stipulate that if the centroid Euclidean distance between the detected clumps and the simulated clumps is not greater than two pixels (The minimum standard deviation of Gaussian distribution), the clumps are regarded as detected.

The 100 second-Galactic-quadrant synthesized data are intercepted along the velocity direction, yielding 42,400 slices of size 361 × 241. The slices are inputted into the trained MCD-YOLOv5 to obtain the detection results. During MCD-YOLOv5 detection, the confidence is set to 0.6, i.e., the location information of the region is recorded only when the probability of being a clump exceeds 0.6. Then, the detection results of the same synthesized data are inputted into DPC to obtain the detection results of clumps in the synthesized data. Using MCD-YOLOv5 joint DPC, 9841 simulated clumps are correctly detected by matching with the simulated clump information tables, with a recall rate of 98.41%. Figure 12 shows the velocity-integrated intensity map of a detection result, where the white circles are the center of mass positions of the simulated clumps, and the red dots indicate the center of mass positions of the clumps that are detected and matched with the simulated clumps by MCD-YOLOv5 joint DPC. Figure 13 shows the integrated intensity map of the detected clump of second-quadrant synthesized data of different directions, the maximum spectrum, and the average spectrum. As can be seen from the figure, the area detected by MCD-YOLOv5 joint DPC matches the characteristics of the simulated clumps.

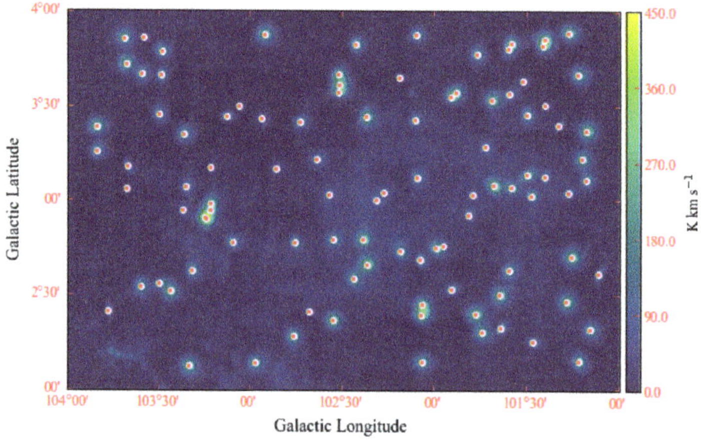

Figure 12. Velocity−integrated intensity maps of the detection result. The white circles are the center of mass positions of the simulated clumps, and the red dots indicate the center of mass positions of the clumps that are detected and matched with the simulated clumps by MCD-YOLOv5 joint DPC.

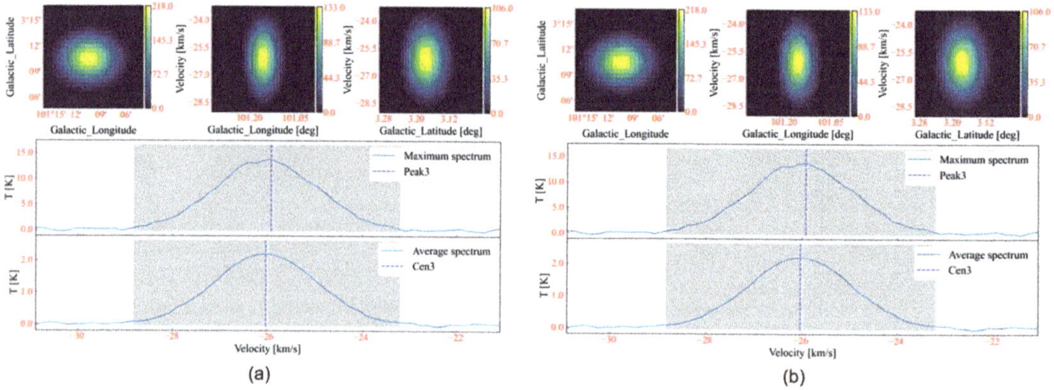

Figure 13. The information of the simulated clump is shown. (**a**,**b**) two examples of the integrated intensity map of the detected clumps of second quadrant synthesized data. The top three subplots offer the l−b, l−v, and b−v maps integration in three directions for a clump, while the middle and bottom of the figure show the peak spectrum and average spectrum of the clump, respectively.

To compare the performance of MCD-YOLOv5 joint DPC with the traditional clump-detection algorithms, we use FellWalker and ClumpFind on the same synthesized data. To improve the detection performance of FellWalker and ClumpFind, both algorithms are parameter-tuned according to the synthesized data to achieve better performance before detection. The detection parameters are determined by the value of the RMS of the MWISP and the experience of the users. When using MCD-YOLOv5 joint DPC, only two parameters need to be used, the threshold of density and distance in DPC. The detection parameters of FellWalker, ClumpFind, and MCD-YOLOv5 joint DPC are shown in Tables 5, 6, and 7, respectively. We determine whether each simulated clump is detected by matching the clump parameter tables with the output center-of-mass information of all three algorithms. Based on statistical results, 9841 clumps were detected by MCD-YOLOv5 joint DPC, 9770 by FellWalker, and 9631 by ClumpFind. When the second-Galactic-quadrant synthesized data was detected using MCD-YOLOv5 joint DPC, 10,584 clumps were detected. Since the synthesized data includes the observational data background and simulated clumps, it needs to consider the number of clumps in the background data when calculating the error rate. The remaining 743 detected clumps were manually identified by manual verification, 495 clumps were falsely detected, and the false detection rate was 4.68%. Manual verification is based on the integrated intensity maps of the detected clump in different directions.

Table 5. FellWalker Parameters.

Parameters Name And Default Value
FELLWALKER.ALLOWEDGE = 1
FELLWALKER.CLEANITER = 1
FELLWALKER.FLATSLOPE = 2×RMS
FELLWALKER.FWHMBEAM = 2
FELLWALKER.MAXBAD = 0.05
FELLWALKER.MAXJUMP = 4
FELLWALKER.MINDIP = 1×RMS
FELLWALKER.MINHEIGHT = 3×RMS
FELLWALKER.MINPIX = 27
FELLWALKER.NOISE = 2×RMS
FELLWALKER.VELORES = 2

Table 6. ClumpFind parameters.

Parameters Name And Default Value
CLUMPFIND.ALLOWEDGE = 1
CLUMPFIND.DELTAT = 2×RMS
CLUMPFIND.FWHMBEAM = 2
CLUMPFIND.IDLAIG = 1
CLUMPFIND.MAXBAD = 0.05
CLUMPFIND.MINPIX = 27
CLUMPFIND.NAXIS = 3
CLUMPFIND.NOISE = 2×RMS
CLUMPFIND.TLOW = 3×RMS
CLUMPFIND.VELORES = 2

Table 7. MCD-YOLOv5 joint DPC parameters. minRho represents the minimum peak intensity of the clump, i.e., the point corresponding to this intensity can be used as the cluster center during the DPC clustering process. minRho can be set according to the intensity characteristics of the clumps in different regions. minDelta represents the minimum pixel distance to distinguish between two clumps. minDelta can be set according to the sparseness of the clump distribution.

Parameters Name	Explanation	Default Value
minRho	The minimum intensity of clump	[2, 5]×RMS
minDelta	The minimum pixel distance to distinguish between two clumps	4

In terms of the total number of detections, the method in this paper is slightly higher, but the number of parameters is greatly reduced and the setup is simpler. The recall rates based on MCD-YOLOv5 joint DPC, FellWalker, and ClumpFind detection results are shown in Table 8. Figure 14 demonstrates the variation of recall with flux and Peak Signal-to-Noise Ratio (PSNR). PSNR and flux of each simulated clump can be calculated by the clump parameter table. After the detection is completed, the detection results of the clumps are matched with the simulated clump parameter table. Clumps in the detection results that satisfy the matching rule are recorded as clumps detected correctly. The recall rate is statistically plotted according to the different intervals of PSNR and flux, which can show the detection performance of the algorithm for different PSNR and flux of clumps. It can be seen that MCD-YOLOv5 joint DPC is higher than the other two methods in most positions of the performance curve. In other words, the method in this paper requires only a small number of parameters to achieve the detection performance of the traditional algorithm with optimized parameters.

Table 8. Detection results of MCD-YOLOv5 joint DPC, FellWalker, and ClumpFind.

Method	Matched Clumps	Recall
MCD-YOLOv5 joint DPC	9841	98.41%
FellWalker	9770	97.70%
ClumpFind	9631	96.31%

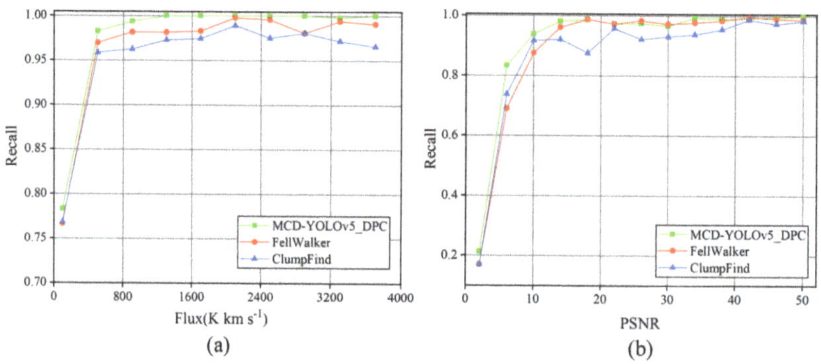

Figure 14. Variation of recall with flux and PSNR for ClumpFind, FellWalker, and MCD-YOLOv5 joint DPC: (**a**) Recall variation with flux, and (**b**) Recall variation with PSNR.

4.4. Detection Results of Observational Data

To verify the feasibility of MCD-YOLOv5 joint DPC on the observational data, we selected a region in the third Galactic quadrant of the ^{13}CO(1-0) line emission data obtained by MWISP for testing. The region selected ranges $180° < l < 195°$ and $-5° < b < 5°$, and the velocity ranges -200 km s^{-1} $< vs. <$ 200 km s^{-1}. This region is divided into different data blocks at 1° grid spacing corresponding with the size of $121 \times 121 \times 2411$ pixels. To obtain annotation information, we used the LDC algorithm to detect this region and obtained the clumps and their corresponding parameter information. We selected the clumps among 200 data blocks with 487 clumps to obtain 8496 samples and their annotation information by the method mentioned in Section 4.2.1. The pixel size of the sample is 121×121. Figure 15 shows the examples from the observational dataset and the annotation information on the samples. The dataset is divided in the same way as in Section 4.2.2. The training parameter is the default parameter of YOLOv5. To reduce the complexity of model training, we used Transfer Learning [35] when training MCD-YOLOv5 with the observational dataset. The trained model parameters on the synthesized dataset in Section 4.2.2 are used as the pre-training weights for the model training on the observational dataset. Figure 16 shows variation curves of loss-change, precision, and recall on the validation set of the observational dataset. Figure 16a,b shows that MCD-YOLOv5 is not overfitted or underfitted when trained on the observational dataset. We selected some samples in the validation set and inputted them into the trained MCD-YOLOv5 for detection.

The detection results of MCD-YOLOv5 are shown in Figure 17. It can be seen that MCD-YOLOv5 can obtain the correct detection position of clump candidates in the 2D PP plane. After MCD-YOLOv5, we use DPC to obtain the final detection results of clumps in PPV space. The detection process and result outputs for observational data are identical to that for synthetic data, which is described in Section 4.3. Figure 18 demonstrates the velocity-integrated intensity maps of some of the detection results. It can be seen that the clumps are detected in a consistent location using both methods, MCD-YOLOv5 joint DPC and FellWalker. Figure 19 shows the integrated intensity map of the detected clump of observational data of different directions, the maximum spectrum, and the average spectrum. As can be seen from the figure, the area detected by MCD-YOLOv5 joint DPC matches the characteristics of the clumps.

By visual inspection, the detected candidate regions match the empirical criteria, demonstrating the viability of the proposed method in observational data. Moreover, the method in this paper does not detect spurious clumps caused by pure noise because there is no such type of target in the training labels. However, since the proposed method is based on supervised deep learning, the final performance is affected by the number of training samples and the accuracy of labeled information. During the experiments, there were also some missed detections of clumps due to incomplete training sets. When the background intensity is greater than the clump, the background will obscure the clumps, resulting in MCD-YOLOv5 missing the detection during the two-dimensional detection on the Galactic plane in the first stage. At the same time, a Gaussian-type clump occurs in multiple velocity channels, and the area of the clump is relatively small at the start and end of the velocity channel. MCD-YOLOv5 is prone to missed detection. Due to the loss of information at both ends, the resulting clump area will be smaller than the actual clump area after the DPC algorithm clusters. In this case, using the results of clustering to calculate the centroid will cause the result to be skewed from the actual position. Therefore, when compared with the simulated clump parameter table, the distance exceeds the size specified by the matching rule, so it is judged to be missed. In fact, the results of DPC clustering show that this type of clump is detected. In conclusion, the method in this paper is an attempt at deep learning for clump target detection in the PPV space.

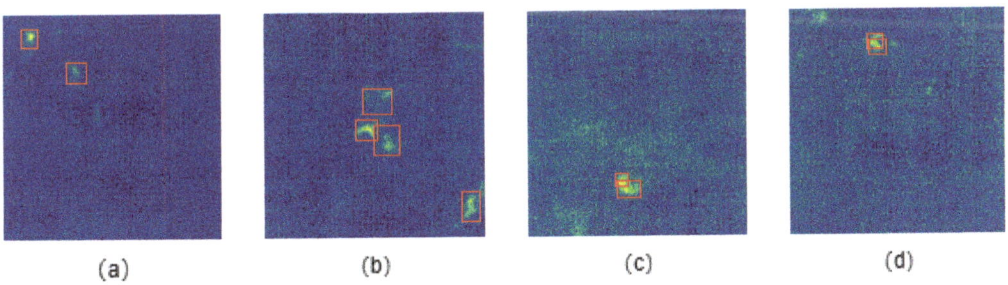

Figure 15. Examples from the MCD-YOLOv5 dataset and labeling the annotation information on the samples. (**a–d**) all show the labeling of the observational data slices in the third Galactic quadrant. The red-labeled box is annotation information.

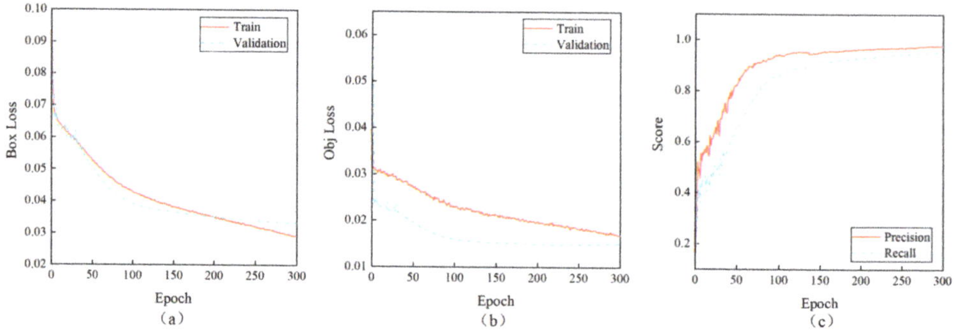

Figure 16. Variation curves of loss-change, precision, and recall on the validation set of the real dataset during MCD-YOLOv5 training: (**a**) variation in localization loss, (**b**) variation in confidence loss, (**c**) variation in precision and recall.

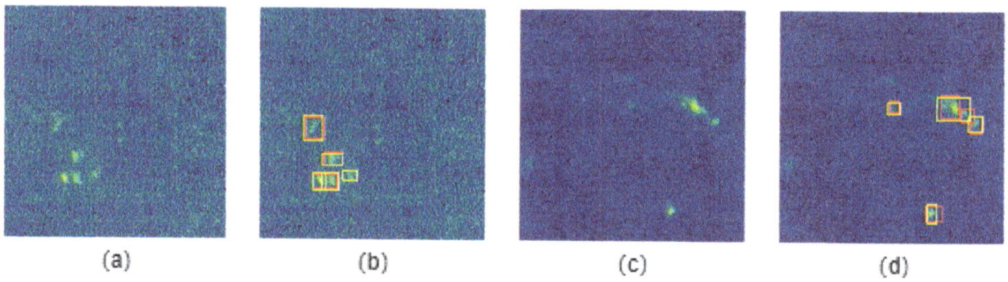

Figure 17. Detection results of MCD-YOLOv5 in observational data. (**a**,**c**) the slices generated by intercepting along the velocity channel of the observational data. (**b**,**d**) the detection results of MCD-YOLOv5. The red-labeled box is annotation information, the yellow-labeled box is the detection result of MCD-YOLOV5.

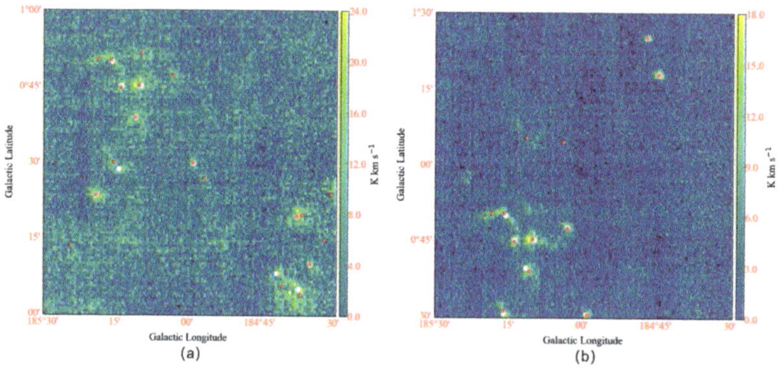

Figure 18. Velocity−integrated intensity maps of detection results. (**a**,**b**) velocity−integrated intensity maps of detection results of the two examples. The white circles are the center of mass positions of the clumps detected by MCD-YOLOv5 joint DPC, and the red dots indicate the center of mass positions of the clumps that are detected by FellWalker.

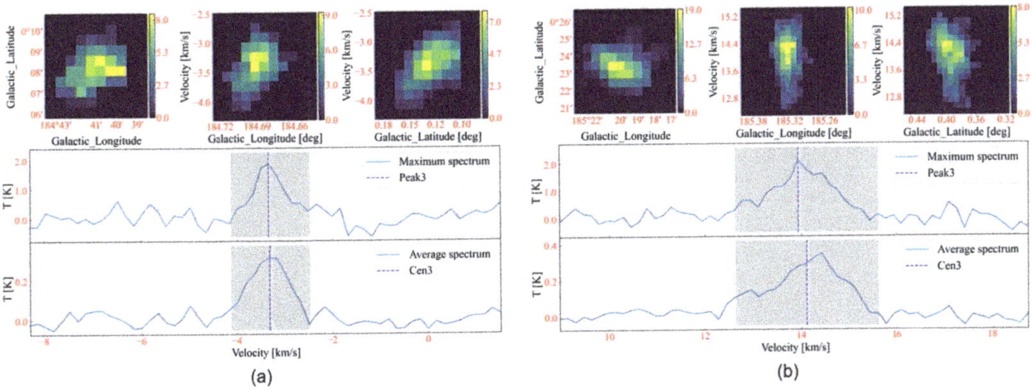

Figure 19. The information of the clump is shown. (**a**,**b**) two examples of the integrated intensity map of the detected clumps of observational data. The top three subplots offer the l−b, l−v, and b−v maps integration in three directions for a clump, while the middle and bottom of the figure show the peak spectrum and average spectrum of the clump, respectively.

5. Summary

We propose a method based on an improved YOLOv5 joint DPC to realize the automatic detection of molecular clumps. Among them, YOLOv5 adds the CA module to the original backbone and modifies the loss function, which is used to improve the detection performance of small clumps. The intensity of the center of mass is considered in the process of DPC and is used to improve the clustering effect of the clumps. The algorithm localizes the molecular clumps on the Galactic plane by using MCD-YOLOv5 and combines the detection results in the velocity direction by using DPC to realize molecular-clump detection in PPV space. We constructed a large amount of synthetic data to test the detection performance of the algorithm. Experimental results show that the method proposed has fewer initial parameters, higher detection efficiency, and achieves the same detection performance as parameter-tuned FellWalker and ClumpFind. From the experiments with observational data, the method in this paper can also accurately find the positions where the clump candidates are located. Subsequently, the generalization ability of the model in observational data can be further improved by optimizing the quality of labels.

Author Contributions: Conceptualization, Z.-W.C. and Y.H.; methodology, S.Z. and Y.H.; software, J.-B.H. and X.-Y.L.; validation, C.L. and X.-Y.L.; formal analysis, J.-B.H. and C.L.; investigation, J.-B.H. and X.-Y.Z.; resources, Z.-W.C. and X.-Y.Z.; data curation, J.-B.H., C.L. and X.-Y.L.; writing—original draft preparation, J.-B.H. and Y.H.; writing—review and editing, X.-Y.Z. and S.Z.; visualization, J.-B.H., and X.-Y.L. All authors have read and agreed to the published version of the manuscript.

Funding: This research was funded by the National Natural Science Foundation of China U2031202.

Institutional Review Board Statement: Not applicable.

Informed Consent Statement: Not applicable.

Data Availability Statement: The testing data that support the study are openly available in https://github.com/SunetK/MCD-YOLOv5-joint-DPC (accessed on 16 August 2023). Other datasets used or analyzed during the study are available from the corresponding author upon reasonable request.

Acknowledgments: We are grateful to the anonymous referee for their invaluable insights and comments, which enabled us to refine and enhance this work. This work is supported by the National Natural Science Foundation of China (grants Nos. U2031202). This research makes use of the data from the Milky Way Imaging Scroll Painting (MWISP) project, which is a multi-line survey in ^{12}CO, ^{13}CO, and C^{18}O along the northern Galactic plane with PMO-13.7m telescope. We are grateful to all the members of the MWISP working group, particularly the staff members at PMO-13.7m telescope, for their long-term support.

Conflicts of Interest: The authors declare no conflicts of interest.

Notes

1 https://github.com/SunetK/MCD-YOLOv5-joint-DPC(accessed on 16 August 2023).

References

1. Heyer, M.; Dame, T.M. Molecular Clouds in the Milky Way. *Annu. Rev. Astron. Astrophys.* **2015**, *53*, 583–629. [CrossRef]
2. Williams, J.P.; Blitz, L.; McKee, C.F. The Structure and Evolution of Molecular Clouds: From Clumps to Cores to the IMF. *arXiv* **1999**, arXiv:astro-ph/9902246.
3. Stutzki, J.; Guesten, R. High Spatial Resolution Isotopic CO and CS Observations of M17 SW: The Clumpy Structure of the Molecular Cloud Core. *Astrophys. J.* **1990**, *356*, 513. [CrossRef]
4. Krumholz, M.R.; McKee, C.F.; Tumlinson, J. The Star Formation Law in Atomic and Molecular Gas. *Astrophys. J.* **2009**, *699*, 850–856. [CrossRef]
5. Zinnecker, H.; Yorke, H.W. Toward Understanding Massive Star Formation. *Annu. Rev. Astron. Astrophys.* **2007**, *45*, 481–563. [CrossRef]
6. Williams, J.P.; de Geus, E.J.; Blitz, L. Determining Structure in Molecular Clouds. *Astrophys. J.* **1994**, *428*, 693. [CrossRef]
7. Berry, D.S. FellWalker-A clump identification algorithm. *Astron. Comput.* **2015**, *10*, 22–31. [CrossRef]
8. Kirk, H.; Di Francesco, J.; Johnstone, D.; Duarte-Cabral, A.; Sadavoy, S.; Hatchell, J.; Mottram, J.C.; Buckle, J.; Berry, D.S.; Broekhoven-Fiene, H.; et al. The JCMT Gould Belt Survey: A First Look at Dense Cores in Orion B. *Astrophys. J.* **2016**, *817*, 167. [CrossRef]
9. Luo, X.; Zheng, S.; Huang, Y.; Zeng, S.; Zeng, X.; Jiang, Z.; Chen, Z. Molecular Clump Extraction Algorithm Based on Local Density Clustering. *Res. Astron. Astrophys.* **2022**, *22*, 015003. [CrossRef]
10. Chen, Q.; Wang, Y.; Yang, T.; Zhang, X.; Cheng, J.; Sun, J. You Only Look One-Level Feature. In Proceedings of the IEEE Conference on Computer Vision and Pattern Recognition, CVPR 2021, Virtual, 19–25 June 2021; pp. 13039–13048. [CrossRef]
11. Wang, J.; Song, L.; Li, Z.; Sun, H.; Sun, J.; Zheng, N. End-to-End Object Detection With Fully Convolutional Network. In Proceedings of the IEEE Conference on Computer Vision and Pattern Recognition, CVPR 2021, Virtual, 19–25 June 2021; pp. 15849–15858. [CrossRef]
12. Yan, B.; Peng, H.; Wu, K.; Wang, D.; Fu, J.; Lu, H. LightTrack: Finding Lightweight Neural Networks for Object Tracking via One-Shot Architecture Search. In Proceedings of the IEEE Conference on Computer Vision and Pattern Recognition, CVPR 2021, Virtual, 19–25 June 2021; pp. 15180–15189. [CrossRef]
13. Kumar, A.; Rawat, Y.S. End-to-End Semi-Supervised Learning for Video Action Detection. In Proceedings of the IEEE/CVF Conference on Computer Vision and Pattern Recognition, CVPR 2022, New Orleans, LA, USA, 18–24 June 2022; pp. 14680–14690. [CrossRef]
14. Liang, C.; Wang, W.; Zhou, T.; Yang, Y. Visual Abductive Reasoning. In Proceedings of the IEEE/CVF Conference on Computer Vision and Pattern Recognition, CVPR 2022, New Orleans, LA, USA, 18–24 June 2022; pp. 15544–15554. [CrossRef]
15. Zhou, K.; Yang, J.; Loy, C.C.; Liu, Z. Conditional Prompt Learning for Vision-Language Models. In Proceedings of the IEEE/CVF Conference on Computer Vision and Pattern Recognition, CVPR 2022, New Orleans, LA, USA, 18–24 June 2022; pp. 16795–16804. [CrossRef]
16. Kim, E.J.; Brunner, R.J. Star-galaxy classification using deep convolutional neural networks. *Mon. Not. R. Astron. Soc.* **2017**, *464*, 4463–4475. [CrossRef]
17. González, R.E.; Muñoz, R.P.; Hernández, C.A. Galaxy detection and identification using deep learning and data augmentation. *Astron. Comput.* **2018**, *25*, 103–109. [CrossRef]
18. Leung, H.W.; Bovy, J. Deep learning of multi-element abundances from high-resolution spectroscopic data. *Mon. Not. R. Astron. Soc.* **2019**, *483*, 3255–3277. [CrossRef]
19. Xie, J.; Bu, Y.; Liang, J.; Li, H.; Wang, X.; Pan, J. Improve the Search of Very Metal-poor Stars Using the Deep Learning Method. *Astron. J.* **2021**, *162*, 155. [CrossRef]
20. He, Z.; Qiu, B.; Luo, A.L.; Shi, J.; Kong, X.; Jiang, X. Deep learning applications based on SDSS photometric data: Detection and classification of sources. *Mon. Not. R. Astron. Soc.* **2021**, *508*, 2039–2052. [CrossRef]
21. Yi, Z.; Li, J.; Du, W.; Liu, M.; Liang, Z.; Xing, Y.; Pan, J.; Bu, Y.; Kong, X.; Wu, H. Automatic detection of low surface brightness galaxies from Sloan Digital Sky Survey images. *Mon. Not. R. Astron. Soc.* **2022**, *513*, 3972–3981. [CrossRef]
22. Cao, Z.; Yi, Z.; Pan, J.; Su, H.; Bu, Y.; Kong, X.; Luo, A. L-dwarf Detection from SDSS Images using Improved Faster R-CNN. *Astron. J.* **2023**, *165*, 184. [CrossRef]
23. Su, Y.; Yang, J.; Zhang, S.; Gong, Y.; Wang, H.; Zhou, X.; Wang, M.; Chen, Z.; Sun, Y.; Chen, X.; et al. The Milky Way Imaging Scroll Painting (MWISP): Project Details and Initial Results from the Galactic Longitudes of 25.°8-49.°7. *Astrophys. J. Suppl. Ser.* **2019**, *240*, 9. [CrossRef]
24. Rodriguez, A.; Laio, A. Clustering by fast search and find of density peaks. *Science* **2014**, *344*, 1492–1496. [CrossRef]
25. Matsubara, T. Analytic Minkowski functionals of the cosmic microwave background: Second-order non-Gaussianity with bispectrum and trispectrum. *Phys. Rev. D* **2010**, *81*, 083505. [CrossRef]
26. Li, Z.; Wang, Y.; Chen, K.; Yu, Z. Channel Pruned YOLOv5-based Deep Learning Approach for Rapid and Accurate Outdoor Obstacles Detection. *arXiv* **2022**, arXiv:2204.13699.

27. Darapaneni, N.; Kumar, S.; Krishnan, S.; Rajagopal, A.; Nagendra.; Paduri, A.R. Implementing a Real-Time, YOLOv5 based Social Distancing Measuring System for COVID-19. *arXiv* **2022**, arXiv:2204.03350.
28. Ewaidat, H.A.; Brag, Y.E. Identification of lung nodules CT scan using YOLOv5 based on convolution neural network. *arXiv* **2022**, arXiv:2301.02166.
29. Jain, S. Adversarial Attack on Yolov5 for Traffic and Road Sign Detection. *arXiv* **2023**, arXiv:2306.0607.
30. Hou, Q.; Zhou, D.; Feng, J. Coordinate Attention for Efficient Mobile Network Design. In Proceedings of the 2021 IEEE/CVF Conference on Computer Vision and Pattern Recognition (CVPR), Nashville, TN, USA, 20–25 June 2021.
31. Yu, J.; Jiang, Y.; Wang, Z.; Cao, Z.; Huang, T. UnitBox: An Advanced Object Detection Network. *arXiv* **2016**, arXiv:1608.01471.
32. Wang, J.; Xu, C.; Yang, W.; Yu, L. A Normalized Gaussian Wasserstein Distance for Tiny Object Detection. *arXiv* **2022**, arXiv:2110.13389.
33. Zheng, Z.; Wang, P.; Liu, W.; Li, J.; Ye, R.; Ren, D. Distance-IoU Loss: Faster and Better Learning for Bounding Box Regression. *arXiv* **2019**, arXiv:1911.08287.
34. Otsu, N. A Threshold Selection Method from Gray-Level Histograms. *IEEE Trans. Syst. Man, Cybern.* **1979**, *9*, 62–66. [CrossRef]
35. Pan, S.J.; Yang, Q. A Survey on Transfer Learning. *IEEE Trans. Knowl. Data Eng.* **2010**, *22*, 1345–1359. [CrossRef]

Disclaimer/Publisher's Note: The statements, opinions and data contained in all publications are solely those of the individual author(s) and contributor(s) and not of MDPI and/or the editor(s). MDPI and/or the editor(s) disclaim responsibility for any injury to people or property resulting from any ideas, methods, instructions or products referred to in the content.

universe

MDPI

Article

Extraction of Physical Parameters of RRab Variables Using Neural Network Based Interpolator

Nitesh Kumar [1,*,†], Harinder P. Singh [2,†], Oleg Malkov [3,†], Santosh Joshi [4], Kefeng Tan [5], Philippe Prugniel [6] and Anupam Bhardwaj [7]

1 Department of Physics, Applied Science Cluster, University of Petroleum and Energy Studies (UPES), Dehradun 248007, India
2 Department of Physics & Astrophysics, University of Delhi, Delhi 110007, India; hpsingh@physics.du.ac.in
3 Institute of Astronomy of the Russian Academy of Sciences (INASAN), 48 Pyatnitskaya St., Moscow 119017, Russia; malkov@inasan.ru
4 Aryabhatta Research Institute of Observational Sciences (ARIES), Manora Peak, Nainital 263001, India; santosh@aries.res.in
5 National Astronomical Observatories, Chinese Academy of Sciences (NAOC), 20A Datun Road, Chaoyang District, Beijing 100101, China; tan@nao.cas.cn
6 Centre de Recherche Astrophysique de Lyon (CRAL), Observatoire de Lyon, 9 Avenue Charles André, 69230 Saint-Genis-Laval, France; prugniel@obs.univ-lyon1.fr
7 Inter-University Centre for Astronomy and Astrophysics (IUCAA), Post Bag 4, Ganeshkhind, Pune 411007, India
* Correspondence: nitesh.kumar@ddn.upes.ac.in
† These authors contributed equally to this work.

Abstract: Determining the physical parameters of pulsating variable stars such as RR Lyrae is essential for understanding their internal structure, pulsation mechanisms, and evolutionary state. In this study, we present a machine learning framework that uses feedforward artificial neural networks (ANNs) to infer stellar parameters—mass (M), luminosity ($\log(L/L_{\odot})$), effective temperature ($\log(T_{\rm eff})$), and metallicity (Z)—directly from Transiting Exoplanet Survey Satellite (TESS) light curves. The network is trained on a synthetic grid of RRab light curves generated from hydrodynamical pulsation models spanning a broad range of physical parameters. We validate the model using synthetic self-inversion tests and demonstrate that the ANN accurately recovers the input parameters with minimal bias. We then apply the trained model to RRab stars observed by the TESS. The observed light curves are phase-folded, corrected for extinction, and passed through the ANN to derive physical parameters. Based on these results, we construct an empirical period–luminosity–metallicity (PLZ) relation: $\log(L/L_{\odot}) = (1.458 \pm 0.028)\log(P/\text{days}) + (-0.068 \pm 0.007)\,[\text{Fe/H}] + (2.040 \pm 0.007)$. This work shows that ANN-based light-curve inversion offers an alternative method for extracting stellar parameters from single-band photometry. The approach can be extended to other classes of pulsators such as Cepheids and Miras.

Keywords: stellar parameters; RR Lyrae variables; neural network

1. Introduction

RR Lyrae stars are low-mass ($0.5 \lesssim M/M_{\odot} \lesssim 0.8$), evolved stellar objects (age $\gtrsim$ 10 Gyr [1]) that occupy the intersection of the horizontal branch and the classical instability strip in the Hertzsprung–Russell diagram. These stars are undergoing a core helium-burning phase, which is similar to the evolutionary stage of intermediate-mass classical Cepheids ($3 \lesssim M/M_{\odot} \lesssim 13$). Owing to their well-established period–luminosity relations (PLRs) in the infrared regime—originally identified by Longmore et al. [2] and subsequently refined by several studies (e.g., [3–7])—RR Lyrae variables serve as reliable distance indicators and are critical to calibrating the cosmic distance ladder [8,9]. Furthermore, these stars offer important insights into stellar evolution and pulsation physics [10], and they are effective tracers of ancient stellar populations in various galactic environments [11].

The physical parameters of RR Lyrae stars are traditionally estimated using empirical relations that connect various light curve characteristics—such as Fourier decomposition parameters and color indices—with the pulsation period [12–14]. However, advances in the theoretical modeling of stellar pulsation have enabled the construction of extensive model grids to investigate the intrinsic properties of RR Lyrae and other variable stars [15–17]. Notably, the radial stellar pulsation (RSP) module developed by Smolec and Moskalik [18], implemented within the Modules for Experiments in Stellar Astrophysics framework (`MESA`; [19–23]), offers a powerful tool for generating theoretical pulsation models. These models facilitate the derivation of fundamental stellar parameters—such as mass, luminosity, and effective temperature—thereby providing deeper insights into the structure and evolution of RR Lyrae stars.

The motivation for inferring fundamental physical parameters—such as mass, luminosity, effective temperature, and metallicity—from light curves comes from the proliferation of time-domain photometric data from large-scale surveys. While double-lined eclipsing binaries and resolved spectroscopic binaries provide the most precise parameter estimates, such systems are rare with only a few hundred thoroughly characterized examples [24,25]. In contrast, wide-field photometric surveys like EROS [26], MACHO [27], OGLE [28], ASAS [29], TrES [30], HAT [31], CoRoT [32], and Kepler [33] have yielded millions of high-quality light curves for pulsating variables, including RR Lyrae stars. This unprecedented data volume makes light curve-based inference methods increasingly attractive for characterizing stellar populations on a large scale, particularly where spectroscopic data are lacking.

An alternative method for inferring the physical parameters of RR Lyrae stars involves directly comparing observed light curves with a reference library of theoretical models. For instance, Das et al. [34] applied this approach to a sample of RRab stars in the Large Magellanic Cloud (LMC), estimating their physical parameters by matching observed light curves with those from the model grid of Marconi et al. [15]. However, this technique is constrained by the limited coverage of the model grid, as only a small subset of observed LMC light curves closely matched the available theoretical templates. A similar technique is used by Kumar et al. [35] for deriving physical parameters of stars in a globular cluster using medium resolution spectra.

A more sophisticated strategy involves constructing a denser and smoother model grid and employing non-linear optimization techniques for parameter estimation [36]. Nevertheless, generating such a comprehensive model grid is computationally demanding, posing practical limitations. To overcome this challenge, Kumar et al. [37] developed an artificial neural network (ANN) trained on the model grid of Marconi et al. [15] for RRab stars in the V and I photometric bands. The resulting ANN[1] serves as an efficient interpolator that is capable of producing high-resolution, smooth model light curves in approximately 55 ms per sample, thereby significantly accelerating the parameter inference process. Hence, we can increase the density of the models using this RRab interpolator. However, the RRab interpolator reliably generates the light curves within the parameter space on which it was trained (see Kumar et al. [37]).

Artificial neural networks (ANNs) offer several advantages in the modeling and inference of pulsating variable star light curves. Once trained, ANNs can perform parameter inference on observed light curves within milliseconds, which is significantly faster than traditional methods such as grid-based forward modeling or interpolation using physical templates [37–39]. Furthermore, ANNs allow the generation of a denser and smoother model grid across the parameter space, improving the fidelity and resolution of the inferred parameters. In our case, the ANN-based interpolator enables efficient mapping from physical parameters to light curves, and vice versa, over a finely sampled synthetic grid. While the current work focuses on the I band, which closely resembles the TESS passband, the approach is generalizable to other photometric bands provided suitable training data are available. These benefits make ANNs a powerful alternative to classical interpolation schemes in the context of variable star analysis.

In this study, we generated a smooth grid of model light curves in the I band over a given parameter space and then trained a reverse interpolator as discussed in Kumar et al. [39] to obtain the physical parameters of non-Blazhko fundamental mode RR Lyrae stars observed in the Transiting Exoplanet Survey Satellite (TESS) field. The I band model light curves were preferred over the V band for training the reverse interpolator, as the TESS light curves exhibit greater similarity to the I band, thereby enabling more accurate parameter estimation from TESS observations.

2. Data and Theoretical Grid

2.1. Synthetic Grid Construction

To accurately determine the physical parameters of pulsating variable stars, it is necessary to compare observed light curves with a grid of model light curves. However, pre-computed grids of hydrodynamical models are typically coarse and unevenly distributed across parameter space, owing to the high computational cost and time required to solve the time-dependent equations governing stellar pulsations. Furthermore, constraints on parameters such as mass, surface gravity, and metallicity often rely on spectroscopic measurements, which are not always available for photometric datasets. Consequently, constructing a fine, dense grid of models becomes essential for reliably inferring stellar properties.

In this work, we generated a fine synthetic grid of RRab light-curve templates in the I band using the RRab interpolator trained by Kumar et al. [37]. We adopted a finer and more uniform sampling where mass (M) varies from 0.5 to 0.8 $M_\odot$ with a constant step size of 0.05 $M_\odot$, luminosity ($\log(L/L_\odot)$) spans from 1.50 to 2.00 dex with a step size of 0.0555 dex, and effective temperature ($T_{\rm eff}$) ranges from 5000 K to 8000 K with a step size of approximately 88.23 K. The metallicity (Z) covers the range from 10^{-4} to 10^{-2} in seven logarithmic steps. The adopted parameter boundaries correspond to the limits of the original training parameter space of the interpolator, and the step sizes were selected to ensure uniform coverage across each parameter dimension. For each value of metallicity, the hydrogen abundance (X) is computed using the relation $X = 1 - Y - Z$, assuming a fixed helium fraction $Y = 0.245$. The parameter space of the synthetic grid is shown in Table 1. Using the trained interpolator, we generate synthetic template light curves in the I band for a grid comprising 17,150 distinct combinations of stellar parameters.

Table 1. The synthetic model light curves are generated in the I band using the RRab interpolator for the given combination of the physical parameters.

Parameter	Range	Step Size
Mass ($M/M_\odot$)	0.5–0.8	0.05
$\log(L/L_\odot)$	1.50–2.00	0.0555
$T_{\rm eff}$ (K)	5000–8000	88.23
Metallicity (Z)	10^{-4}–10^{-2}	logarithmic steps
Hydrogen fraction (X)	0.755–0.745	computed as $1 - Y - Z$

The pulsation period (P) of an RR Lyrae star is closely linked to its mass, luminosity, and effective temperature, as originally described by the van Albada–Baker (vAB) relation [40]. We employ a modern version of the vAB relation, incorporating the dependence on metallicity, as formulated by Marconi et al. [15], to compute the periods for our synthetic models. This relation is specifically calibrated for fundamental-mode (RRab) pulsators.

2.2. Observational TESS Sample

The Transiting Exoplanet Survey Satellite (TESS) is a NASA Astrophysics Explorer mission employing four wide-field CCD cameras to perform a nearly all-sky photometric survey, providing a pre-selected 2-minute cadence and 30-minute full-frame images (FFIs) covering approximately 2300 deg^2 per sector [41]. The continuous, high-precision optical time-series photometry provided by TESS has revolutionized the study of pulsating variables by offering uniformly sampled, multi-sector light curves free from diurnal gaps [41]. Among its many variable-star discoveries, TESS has observed over a thousand RR Lyrae variables, including hundreds of fundamental-mode RRab pulsators, allowing detailed analyses of their pulsation modes, the incidence and characteristics of the Blazhko effect, and low-amplitude secondary oscillations [42]. Differential-image photometry techniques applied to TESS FFIs have cleanly extracted RRab light curves, which, when combined with Gaia parallaxes, have been used to refine empirical period–luminosity–metallicity relationships and to classify RR Lyrae populations across the sky [42].

We compiled a sample of RRab stars by cross-matching known RRab variables from the SIMBAD database with the TESS Input Catalog version 8.2 [43–45]. The selection criteria required that each star had been observed in at least one TESS sector and had available measurements for at least one of the following parameters: effective temperature ($T_{\rm eff}$), surface gravity ($\log g$), or metallicity ([Fe/H]). This process resulted in a final sample of 71 RRab stars. The spatial distribution of these stars is shown in Figure 1. The light curves of these stars were taken from the TESS archive using lightkurve[2] [46] tool in Python. We converted the raw Pre-search Data Conditioning Simple Aperture Photometry (PDCSAP) flux to TESS magnitude using a zero point magnitude equal to 20.44 from TESS data release notes[3]. We converted the TESS magnitude to I band magnitude by deriving a relation between the TESS magnitude and I band magnitude using the spectra of a typical RRab star.

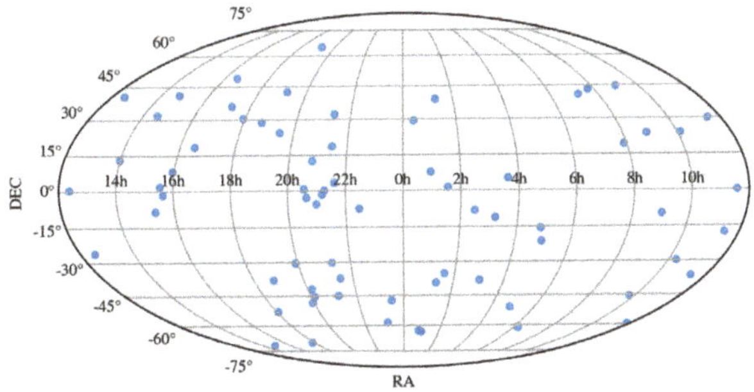

Figure 1. RA(J2000.0)-Dec(J2000.0) The distribution of the 71 RRab stars cross-matched between SIMBAD and the TESS Input Catalog v8.2 [43–45].

3. Parameter Estimator ANN Model

To derive the physical parameters of RRab stars observed by TESS, we first constructed a fine synthetic grid of model light curves covering a broad range of stellar masses, luminosities, effective temperatures, and metallicities. The template light curves were generated in the I band, matching the photometric bandpass most closely related to the

TESS instrumental response. The synthetic grid comprised 17,150 distinct parameter combinations, each corresponding to a unique model light curve generated using the trained RRab interpolator. The uniform and dense coverage of the grid enables robust interpolation across the relevant regions of parameter space and circumvents the limitations imposed by the sparsity of traditional hydrodynamical models. The TESS light curves, corrected for interstellar extinction and folded on their best periods, serve as the observational inputs for the extraction of physical parameters.

The relationship between the observed light curve (y) and the corresponding set of physical parameters (x) can be expressed as a forward mapping function f such that $y \equiv f(x)$. Inverting this mapping—recovering x given y—requires the construction of an inverse function g such that $x \equiv g(y) = f^{-1}(y)$. Artificial neural networks (ANNs) provide a powerful and flexible framework for approximating such inverse mappings, particularly when the underlying functions are continuous and differentiable [47–49]. By training an ANN on the synthetic grid of light curves and associated stellar parameters, we enable the efficient and accurate retrieval of mass, luminosity, effective temperature, and metallicity for observed RRab stars directly from their light-curve morphology.

We trained a reverse interpolator using a synthetic grid of light curves and corresponding physical parameters. In this setup, the reverse interpolator is a feedforward ANN, where the input is the absolute I band light curve, sampled at 500 equally spaced phase points between 0 and 1, and the output consists of the physical parameters—mass ($M/M_{\odot}$), luminosity ($\log(L/L_{\odot})$), effective temperature ($\log(T_{\rm eff})$), and metallicity (Z).

The input layer consists of 500 absolute I band magnitude values, each corresponding to one of the sampled phase points. Since the physical parameters exhibit vastly different numerical ranges, we employed the Robust Scaler method to scale the output values. This approach scales the data using the interquartile range (IQR), making the network less sensitive to outliers and leading to more stable and efficient training. By transforming the physical parameter outputs into a more uniform scale, we ensured that the ANN training could converge more quickly and avoid issues stemming from disparities in the different parameters.

The most crucial part in training any ANN is the choice of hyperparameters of the network, like the number of hidden layers, number of neurons in each hidden layer, activation function, optimization algorithm and learning rate. The training time and the convergence of the ANN depend on these choices of hyperparameters. We used `RandomSearch` tuner from the `KerasTuner` library for conducting a systematic hyperparameter tuning.

We defined the model architecture in such a way that allows the number of hidden layers to vary between one and six. For each hidden layer, the number of units and activation function were treated as tunable parameters, and the output layer was kept without any activation to predict the physical parameters. We chose the `adam` [50] optimizer with the learning rate sampled logarithmically between 10^{-4} and 10^{-2}. All layers were initialized with the `Glorot uniform` initializer. We trained each model for 50 epochs with a batch size of 128, using the mean squared error (MSE) as both the loss function and tuning objective.

The hyperparameter search explored 100 different configurations in total. Table 2 summarizes the hyperparameter grid. The dataset was divided into three parts: 80% for training, 10% for validation, and the remaining 10% for testing. The final model selection was based on the configuration achieving the minimum validation MSE.

We achieved the lowest validation loss with an ANN model comprising *four* hidden layers containing 256, 128, 32, and 8 neurons, respectively. Each hidden layer has the `ReLU` activation function and includes L2 regularization with a strength of 10^{-6} to mitigate overfitting. The output layer consists of four neurons and no activation function, corresponding to the four scaled predicted physical parameters. A schematic diagram of the final ANN architecture is shown in Figure 2.

Table 2. Hyperparameter grid for ANN model tuning.

Hyperparameter	Values
Number of layers	1 to 6
Units per layer	16, 32, 64, 128, 256, 512
Activation function	ReLU, tanh
Learning rate	10^{-4} to 10^{-2} (log-uniform sampling)
Batch size	32 (fixed)
Optimizer	`adam` [50]
Kernel initializer	Glorot Uniform (fixed seed)
Loss function	Mean Squared Error (MSE)

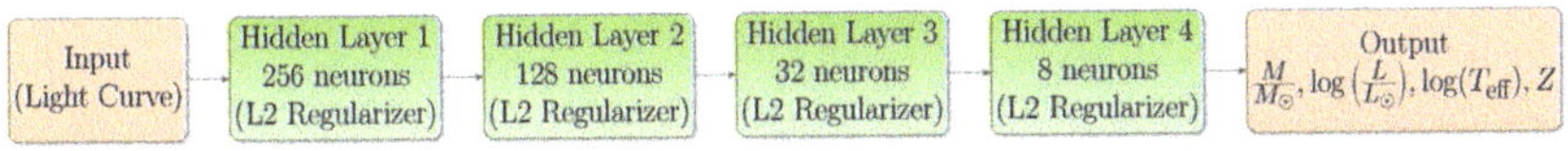

Figure 2. ANN architecture adopted after hyperparameter optimization for physical parameter estimation.

The final ANN model was trained using the `adam` optimizer with an initial learning rate of 0.001 and MSE as the loss function with 80% of the original models used for training, 10% for validation, and the remaining 10% reserved for testing. To facilitate efficient convergence, we implemented a piecewise constant learning rate schedule: the learning rate was decreased by a factor of 5 at batch sizes of 10 and 100, respectively. Training was performed over 1200 epochs with a batch size of 32. The model weights were updated using shuffling at each epoch, and training was parallelized across eight CPU workers with multiprocessing. The total training time was ~27.35 min, utilizing a system equipped with 64 CPU cores operating at a maximum frequency of 3.5 GHz.

The training and validation loss curve for the ANN model, along with the learning rate, is shown in Figure 3. The curve shows the training and validation loss (MSE) during the training over the course of 1200 epochs while also showing the variation in the learning rate. This figure provides insight into both the convergence behavior of the model and the effect of the learning rate.

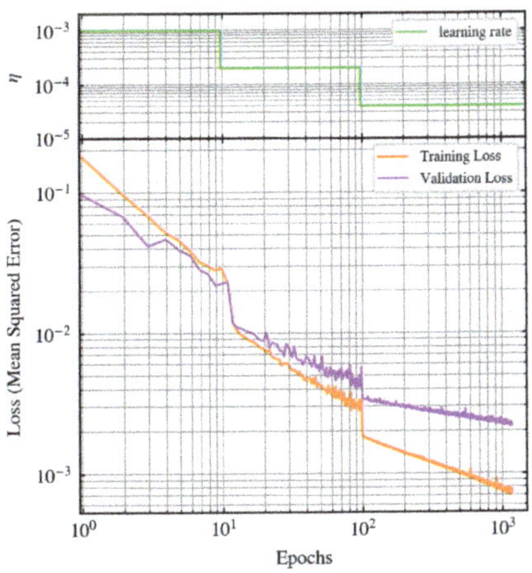

Figure 3. Training and validation loss curves, along with the learning rate progression, for the ANN model over the training epochs.

4. Results

4.1. Self Inversion

To assess the performance of the trained parameter estimator ANN model, we performed a self-inversion test. In this test, the synthetic model I band light curves from the test grid were passed through the trained ANN, and the recovered physical parameters were compared with their original values.

The comparison between the ANN-predicted and true values of mass, luminosity, effective temperature, and metal abundance for the test set is shown in Figure 4. The strong correlation between the predicted and true values demonstrates that the network accurately recovers stellar parameters across the entire grid. The predicted points closely match the original values with negligible scatter or systematic bias, as presented in Table 3.

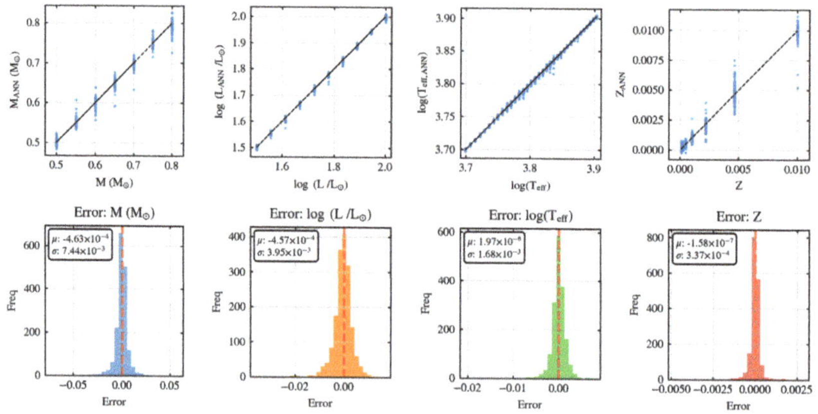

Figure 4. Comparison between the original and predicted values of mass, $\log(L/L_{\odot})$, $\log(T_{\rm eff})$ and Z from the synthetic self-inversion on the test set is shown in top panel. In the bottom panel, the error distribution is shown along with the mean and standard deviation of the error. The close overlap indicates excellent recovery performance of the reverse interpolator for mass, luminosity and temperature.

The relative root-mean-square error (Relative RMSE) provides an intuitive measure of a model's predictive performance across parameters with different numerical scales. It is defined as the root-mean-square error normalized by the mean absolute value of the true parameter values and is expressed as a percentage:

$$\text{Relative RMSE} = \left(\frac{\sqrt{\frac{1}{n}\sum_{i=1}^{n}(x_i - \hat{x}_i)^2}}{\frac{1}{n}\sum_{i=1}^{n}|x_i|} \right) \times 100\%$$

where x_i and $\hat{x}_i$ are the true and predicted values of the parameter, respectively.

Table 3. Performance of the trained parameter estimator model on the test set based on MSE and relative root-mean-square error (Relative RMSE). The Relative RMSE is computed as the RMSE normalized by the mean absolute value of the true parameter values.

Parameter	MSE	Relative RMSE (%)
Mass (M)	5.6×10^{-5}	1.15
Luminosity (L)	1.6×10^{-5}	0.23
$\log(T_{\rm eff})$	3.1×10^{-6}	0.04
Metallicity (Z)	1.1×10^{-7}	12.81

The synthetic self-inversion tests confirm that the trained parameter estimator model is capable of recovering the physical parameters of RRab stars from their light curves with high accuracy and minimal bias. For example, the model achieves relative RMSE values of only 1.15% for mass, 0.23% for luminosity, and 0.04% for $\log(T_{\rm eff})$, indicating excellent agreement between predicted and true values. This test validates the internal consistency and predictive reliability of the trained parameter estimator model, although a relatively higher relative RMSE of 12.81% is observed for metallicity.

4.2. Application to TESS RRab Stars

4.2.1. Light-Curve Processing and Folding

To apply the trained parameter estimator model to real data, we selected a sample of RRab stars observed by the TESS for which literature estimates of fundamental stellar parameters are available in Stassun et al. [45]. The TESS light curves were downloaded in the form of short-cadence (2-min) simple aperture photometry (SAP) flux from SPOC Jenkins et al. [51] pipeline FITS files. The fluxes were converted into apparent magnitudes using the standard TESS zero point (adopted from TESS data release notes[4]):

$$m_{\rm TESS} = -2.5\log_{10}(\text{flux}) + 20.44. \quad (1)$$

The period for each light curve was derived using the Lomb–Scargle [52,53] method and then phase-folded using this derived period. Only positive flux points were retained to avoid contamination. To suppress observational noise and fill missing data segments, we fitted a Fourier series to the folded light curve with 10 Fourier components (see Equation (6) of Kumar et al. [39]). The smoothed light curve was sampled at 1000 phase points and then rebinned to 500 evenly spaced bins between phases 0 and 1. This format matches the structure of the input layer used during ANN training.

4.2.2. Photometric Corrections and Calibration

To convert apparent TESS band magnitudes into absolute I band magnitudes, several corrections were applied. Gaia DR3 parallaxes were used to compute distances, from which the distance modulus μ was derived as

$$\mu = 5\log_{10}(d\,[\text{pc}]) - 5. \quad (2)$$

Extinction values were obtained from the IRSA Galactic Dust Reddening Tool, which returns extinction in V band magnitudes (A_V) based on the star's equatorial coordinates using the reddening maps of the Schlafly and Finkbeiner [54]. The I band extinction A_I was estimated using standard extinction ratios and applied to the apparent TESS magnitudes after converting them to the I band. As TESS does not operate in the standard I band, we employed the empirical transformation:

$$I = m_{\rm TESS} - 0.0695. \quad (3)$$

Thus, the absolute I band magnitude for each star becomes

$$M_I = I - \mu - A_I. \quad (4)$$

4.2.3. ANN-Based Parameter Inference

The preprocessed and rebinned absolute I band light curves were then passed through the trained ANN to predict the underlying stellar parameters. The ANN output includes the stellar mass $M/M_\odot$, logarithmic luminosity $\log(L/L_\odot)$, effective temperature $\log(T_{\rm eff})$, and metallicity Z:

$$(M, \log L, \log T_{\rm eff}, Z) = \mathrm{ANN}(y_{\rm TESS}). \tag{5}$$

The effective temperature can be recovered via inverse logarithmic transformation: $T_{\rm eff} = 10^{\log T_{\rm eff}}$. To derive the iron abundance, we converted Z to [Fe/H] using the relation provided by [55]

$$[\mathrm{Fe/H}] = \log_{10}\left(\frac{Z/X}{Z_\odot/X_\odot}\right), \tag{6}$$

where $Z_\odot = 0.0122$ and $X_\odot = 0.7392$ [56]. The hydrogen abundance X is calculated from $X = 1 - Y - Z$, assuming a fixed helium mass fraction $Y = 0.245$. This value corresponds to the primordial helium abundance, which is suitable for RR Lyrae stars given their old evolutionary status. Helium enrichment in such populations is expected to be minimal. Observational support for this assumption is provided by Marconi and Minniti [57], who found little evidence of helium enhancement among RR Lyrae stars in the Galactic bulge.

Table 4 presents the derived physical parameters for a sample of RRab stars based on their TESS light curves, using our trained ANN-based parameter estimator. For each star, the table lists the TESS ID, parallax from the TESS Input Catalog v8.2 [45], the literature period, and the Lomb–Scargle period measured from the TESS data. The effective temperature ($T_{\rm eff}$), mass (M), luminosity (L), and metallicity (Z) inferred by the ANN are compared against corresponding values from the literature when available. We also report the photometric surface gravity ($\log g$).

Figure 5 shows the comparison between the physical parameters of RR Lyrae stars derived from ANN analysis of TESS light curves and those from literature values. The top row displays one-to-one comparisons with literature values on the x-axis and ANN predictions on the y-axis, while the bottom row shows error distributions. The ANN method produces effective temperatures with a mean error of 143 K ($\sigma = 352$ K), metallicities with a mean error of 0.27 dex ($\sigma = 0.83$ dex), masses with a mean error of $-0.74\,M_\odot$ ($\sigma = 0.13\,M_\odot$), and luminosities with a mean error of $9.63\,L_\odot$ ($\sigma = 11.32\,L_\odot$).

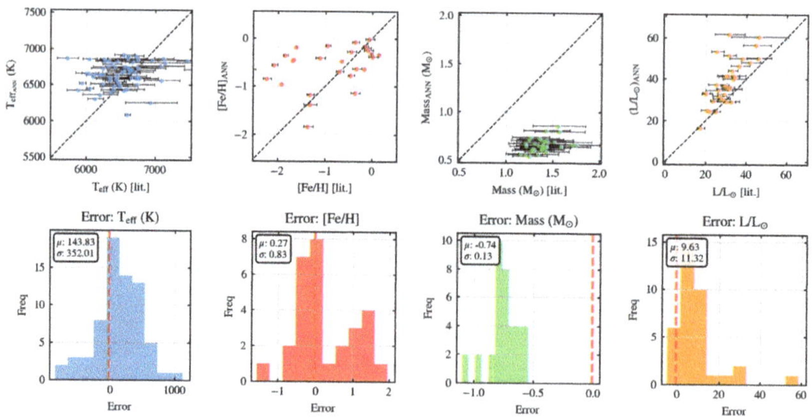

Figure 5. Comparison between literature values and ANN predictions for RR Lyrae stars. **Top row**: One-to-one comparison of literature values (x-axis) versus ANN predictions (y-axis) for effective temperature, metallicity, mass, and luminosity with error bars representing literature uncertainties. **Bottom row**: Distribution of prediction errors (ANN–literature) for each parameter, showing mean (μ) and standard deviation (σ) values. The dashed lines in all panels represent perfect agreement.

Table 4. Physical parameters of RR Lyrae stars derived from TESS light curves using ANN method compared with literature values.

ID	TESS ID	Plx_{lit} (″)	Period (d)	Period LS (d)	$T_{eff_{lit}}$ (K)	$T_{eff_{ANN}}$ (K)	M_{lit} ($M_\odot$)	M_{ANN} ($M_\odot$)	Lum_{lit} ($L_\odot$)	Lum_{ANN} ($L_\odot$)	$[Fe/H]_{lit}$ (dex)	$[Fe/H]_{ANN}$ (dex)	$logg_{lit}$ (dex)
XZ Dra	229913521	1.30 ± 0.02	0.476475	0.476574	6447 ± 224	6824	1.31 ± 0.22	0.67	34.76 ± 1.98	40.91	–	−0.56	3.21 ± 0.12
V1195 Her	232570970	0.53 ± 0.02	0.609729	0.608373	5889 ± 167	6420	–	0.67	–	56.91	−1.60 ± 0.01	−0.19	–
TW Lyn	241174387	0.62 ± 0.04	0.481854	0.481025	6008 ± 162	6659	–	0.62	–	39.10	−0.64 ± 0.01	−0.47	–
TT Lyn	29172806	1.22 ± 0.04	0.597443	0.597684	6191 ± 193	6635	–	0.65	–	52.51	−1.65 ± 0.02	−0.45	–
TW Boo	68076619	0.72 ± 0.02	0.532264	0.531162	6247 ± 42	6645	–	0.70	–	45.32	−1.46 ± 0.07	–	–
SV Eri	9843991	1.22 ± 0.06	0.713900	0.713374	6200 ± 249	6360	–	0.81	–	60.43	−2.06 ± 0.06	−0.56	–
SS Tau	311091712	0.67 ± 0.05	0.369910	0.369768	6724 ± 108	6903	1.43 ± 0.26	0.64	21.14 ± 3.31	25.00	0.17 ± 0.03	−0.33	3.53 ± 0.11
RX Eri	114923989	1.62 ± 0.03	0.587252	0.586150	6443 ± 205	6629	–	0.63	–	49.70	–	−0.56	–
U Lep	146324929	0.93 ± 0.03	0.581458	0.580807	6541 ± 229	6689	–	0.74	–	58.66	–	–	–
TZ Aur	328588068	0.66 ± 0.04	0.391673	0.391437	7128 ± 145	6867	1.59 ± 0.28	0.64	26.79 ± 3.39	31.72	−0.15 ± 0.02	−0.64	3.58 ± 0.10
SZ Gem	63172763	0.59 ± 0.04	0.501165	0.501342	6050 ± 101	6856	–	0.71	–	42.36	−1.65 ± 0.13	–	–
EZ Cnc	197217727	0.50 ± 0.05	0.545777	0.546779	6625 ± 100	6662	1.39 ± 0.23	0.61	31.60 ± 6.34	35.19	−0.00 ± 0.03	−0.32	3.32 ± 0.13
SS Gru	129710422	0.14 ± 0.04	0.959791	0.956326	6942 ± 379	6252	–	0.95	–	291.22	–	−0.56	–
VW Scl	41833926	0.85 ± 0.07	0.510917	0.511665	6672 ± 111	6816	1.41 ± 0.23	0.70	31.06 ± 5.55	41.71	−1.27 ± 0.05	–	3.35 ± 0.11
Z Mic	89358641	0.81 ± 0.07	0.586928	0.585981	6398 ± 200	6480	1.28 ± 0.21	0.60	34.01 ± 5.69	40.01	–	−0.18	3.19 ± 0.13
VX Scl	32282599	0.43 ± 0.03	0.637078	0.635441	6716 ± 208	6597	–	0.76	–	59.59	–	−0.95	–
RZ Cet	1129237	0.68 ± 0.04	0.510600	0.511475	6248 ± 182	6729	1.21 ± 0.18	0.67	25.68 ± 3.12	53.38	−1.90 ± 0.02	−0.96	3.25 ± 0.12
BN Aqr	39999072	0.38 ± 0.06	0.469600	0.468717	6955 ± 760	6849	1.52 ± 0.40	0.69	46.69 ± 14.23	60.60	–	–	3.27 ± 0.31
CP Aqr	248483300	0.78 ± 0.09	0.463400	0.463006	7148 ± 522	6806	1.59 ± 0.32	0.64	26.65 ± 6.66	34.45	–	−0.74	3.59 ± 0.22
SW Aqr	387498678	0.76 ± 0.23	0.459305	0.459871	–	6824	–	0.72	–	43.47	–	–	–
RR Cet	344299442	1.52 ± 0.08	0.553041	0.552368	6650 ± 139	6589	1.40 ± 0.24	0.66	42.32 ± 4.75	48.09	−1.35 ± 0.10	−1.84	3.20 ± 0.10
SX Aqr	353029655	0.62 ± 0.07	0.535708	0.536120	6325 ± 194	6838	1.25 ± 0.19	0.71	31.85 ± 7.46	61.80	–	−1.84	3.19 ± 0.14
AO Peg	283289876	0.35 ± 0.04	0.547245	0.546917	6342 ± 71	6554	–	0.67	–	51.16	−1.25 ± 0.07	−2.81	–
SW And	437761208	1.78 ± 0.16	0.442261	0.441881	6735 ± 138	6613	1.44 ± 0.24	0.59	30.35 ± 6.28	29.13	−0.07 ± 0.10	−0.25	3.38 ± 0.13
DM Cyg	117638854	0.97 ± 0.05	0.419868	0.419951	6415 ± 151	6748	1.29 ± 0.19	0.61	20.62 ± 2.50	32.85	0.03 ± 0.10	−0.38	3.42 ± 0.10
XX And	186452465	0.69 ± 0.05	0.722772	0.721256	6097 ± 9	6613	–	0.70	–	67.68	−1.67 ± 0.01	−0.46	–
DH Hya	47291018	0.47 ± 0.04	0.488996	0.488922	6259 ± 54	6856	–	0.71	–	59.94	−1.52 ± 0.09	–	–
RR Leo	3941985	1.00 ± 0.09	0.452403	0.451699	6593 ± 291	6863	1.37 ± 0.25	0.72	33.79 ± 6.00	41.59	–	–	3.28 ± 0.16
WY Ant	168276785	0.91 ± 0.06	0.574350	0.575440	6765 ± 277	6583	1.45 ± 0.27	0.69	46.47 ± 6.23	50.15	–	–	3.21 ± 0.15

Table 4. *Cont.*

ID	TESS ID	Plx_{lit} (″)	Period (d)	Period LS (d)	$T_{eff_{lit}}$ (K)	$T_{eff_{ANN}}$ (K)	M_{lit} ($M_\odot$)	M_{ANN} ($M_\odot$)	Lum_{lit} ($L_\odot$)	Lum_{ANN} ($L_\odot$)	$[Fe/H]_{lit}$ (dex)	$[Fe/H]_{ANN}$ (dex)	$logg_{lit}$ (dex)
V595 Cen	152231997	0.44 ± 0.03	0.690994	0.691325	6430 ± 191	6569	–	0.65	–	53.30	–	−0.25	–
TU UMa	144376546	1.56 ± 0.06	0.557658	0.556678	6200 ± 65	6643	–	0.65	–	44.82	−1.31 ± 0.14	−1.39	–
SS Leo	49417864	1.13 ± 0.72	0.626351	0.626993	–	6676	–	0.70	–	58.65	–	−1.80	–
W Crt	219249305	0.75 ± 0.04	0.412012	0.412036	6740 ± 305	6845	1.44 ± 0.27	0.64	29.59 ± 3.80	32.68	–	−0.70	3.39 ± 0.15
UV Vir	377172421	0.56 ± 0.05	0.587065	0.586541	7550 ± 130	6714	1.75 ± 0.28	0.66	38.59 ± 6.88	49.96	−1.10 ± 0.10	−0.41	3.56 ± 0.11
UZ CVn	376689735	0.51 ± 0.03	0.697793	0.696838	6329 ± 58	6599	–	0.66	–	56.78	−2.22 ± 0.09	−0.85	–
SV Hya	453469791	1.21 ± 0.05	0.478527	0.478892	6744 ± 324	6850	1.44 ± 0.28	0.71	31.79 ± 3.08	51.43	–	–	3.36 ± 0.15
UY Boo	458457951	0.62 ± 0.05	0.650800	0.651139	6433 ± 292	6581	–	0.77	–	69.01	–	-0.73	–
RS Boo	409373422	1.36 ± 0.04	0.377365	0.377471	6610 ± 132	6719	1.38 ± 0.22	0.61	24.05 ± 1.74	29.52	−0.12 ± 0.10	−0.21	3.43 ± 0.09
V413 CrA	253708643	1.13 ± 0.05	0.589343	0.589858	5945 ± 47	6521	–	0.60	–	50.39	−0.94 ± 0.05	−0.15	–
V4424 Sgr	271404999	1.70 ± 0.04	0.424503	0.424840	6240 ± 181	6511	1.21 ± 0.18	0.55	25.46 ± 1.63	24.36	–	−0.00	3.25 ± 0.10
TV Lib	79403057	0.79 ± 0.04	0.269624	0.269903	6620 ± 132	6897	1.39 ± 0.23	0.64	17.68 ± 1.81	16.29	−0.43 ± 0.10	−0.28	3.57 ± 0.10
BT Aqr	248815852	0.56 ± 0.05	0.406357	0.406222	6700 ± 268	6681	1.42 ± 0.27	0.62	25.69 ± 4.50	37.35	–	−0.55	3.44 ± 0.15
AA Aql	248097218	0.68 ± 0.05	0.361768	0.361829	6550 ± 146	6915	1.35 ± 0.24	0.65	27.71 ± 4.25	30.46	−0.32 ± 0.10	−0.63	3.35 ± 0.12
V456 Ser	46253107	0.22 ± 0.04	0.517558	0.517328	6600 ± 40	6084	–	0.76	–	151.05	−2.64 ± 0.17	−0.66	–
VY Ser	371355048	1.32 ± 0.07	0.714096	0.712959	6055 ± 50	6448	–	0.70	–	64.02	−1.82 ± 0.05	−0.36	–
V341 Aql	375822497	0.84 ± 0.05	0.578042	0.579124	6504 ± 272	6666	–	0.73	–	63.17	–	−1.63	–
AT Ser	264584937	0.58 ± 0.05	0.746609	0.745240	6591 ± 249	6436	–	0.75	–	73.06	–	−0.47	–
DX Del	85211589	1.68 ± 0.03	0.472611	0.472674	6454 ± 118	6534	1.31 ± 0.21	0.60	32.36 ± 1.84	36.11	−0.14 ± 0.06	−0.16	3.24 ± 0.09
VX Her	356085581	0.98 ± 0.06	0.455356	0.456175	6369 ± 135	6786	1.27 ± 0.20	0.74	37.00 ± 4.74	46.23	−1.45 ± 0.07	–	3.14 ± 0.11
BN Vul	111443222	1.40 ± 0.03	0.594132	0.593941	6318 ± 222	6749	–	0.85	–	143.31	–	−0.86	–
TW Her	10581605	0.86 ± 0.02	0.399596	0.400107	7465 ± 147	6836	1.71 ± 0.29	0.66	29.37 ± 1.94	36.89	−0.35 ± 0.10	−1.14	3.65 ± 0.09
CN Lyr	317170031	1.11 ± 0.03	0.411382	0.411798	6355 ± 106	6393	1.26 ± 0.19	0.54	22.33 ± 1.79	24.38	−0.04 ± 0.10	−0.01	3.36 ± 0.09
VZ Her	85516380	0.63 ± 0.02	0.440360	0.440466	5732 ± 153	6870	–	0.71	–	42.43	−1.71 ± 0.01	–	–
FN Lyr	158553034	0.30 ± 0.03	0.527390	0.527304	6745 ± 410	6735	–	0.78	–	77.00	–	−1.09	–
SW For	175492625	0.36 ± 0.02	0.803760	0.803760	6389 ± 176	6426	–	0.67	–	59.24	–	−0.26	–
U Pic	259590223	0.73 ± 0.02	0.440373	0.440466	6642 ± 106	6718	1.40 ± 0.23	0.63	30.57 ± 2.50	35.96	−0.69 ± 0.08	−0.69	3.34 ± 0.09
CD Vel	34069197	0.57 ± 0.03	0.573516	0.572059	7099 ± 261	6558	1.58 ± 0.27	0.80	54.15 ± 6.25	75.49	–	−0.74	3.26 ± 0.12
ST Pic	150166721	2.00 ± 0.02	0.485749	0.486130	6549 ± 182	6422	1.35 ± 0.22	0.56	33.05 ± 1.70	28.96	–	−0.06	3.27 ± 0.10

Table 4. *Cont.*

ID	TESS ID	Plx_{lit} (″)	Period (d)	Period LS (d)	$T_{eff_{lit}}$ (K)	$T_{eff_{ANN}}$ (K)	M_{lit} ($M_\odot$)	M_{ANN} ($M_\odot$)	Lum_{lit} ($L_\odot$)	Lum_{ANN} ($L_\odot$)	$[Fe/H]_{lit}$ (dex)	$[Fe/H]_{ANN}$ (dex)	$logg_{lit}$ (dex)
AE Tuc	234507163	0.63 ± 0.02	0.414528	0.414589	6274 ± 116	6826	1.22 ± 0.17	0.64	20.32 ± 1.79	33.14	−0.45 ± 0.07	−0.78	3.36 ± 0.09
W Tuc	234518883	0.57 ± 0.03	0.642247	0.643122	6042 ± 106	6655	–	0.72	–	66.97	−1.31 ± 0.08	−1.18	–
BI Cen	267930751	0.69 ± 0.03	0.453193	0.452948	7085 ± 341	6765	1.57 ± 0.28	0.86	47.25 ± 5.75	106.42	–	−0.82	3.32 ± 0.14
YY Tuc	220512467	0.43 ± 0.03	0.634857	0.634866	6350 ± 233	6595	–	0.77	–	69.81	–	−1.37	–
RV Phe	425863844	0.52 ± 0.04	0.596419	0.596355	6471 ± 195	6504	–	0.68	–	56.92	–	−0.39	–
TY Aps	258812822	0.62 ± 0.03	0.501720	0.501070	6588 ± 283	6749	–	0.68	–	42.88	–	–	–
EX Aps	294832702	0.58 ± 0.03	0.471799	0.472235	6738 ± 297	6741	1.44 ± 0.27	0.64	29.83 ± 3.51	41.24	–	−0.87	3.39 ± 0.14
V Ind	126910093	1.50 ± 0.04	0.479594	0.479773	6551 ± 105	6742	1.36 ± 0.21	0.73	33.35 ± 2.39	44.11	−1.32 ± 0.03	–	3.27 ± 0.08
MS Ara	337440887	0.55 ± 0.04	0.524988	0.525991	6735 ± 254	6503	1.44 ± 0.25	0.81	45.39 ± 6.95	56.61	–	−0.69	3.21 ± 0.13
CD-45 13628	129081118	1.26 ± 0.04	0.510247	0.510016	6133 ± 115	6305	1.16 ± 0.16	0.56	28.45 ± 2.40	30.13	−0.35 ± 0.08	−0.08	3.15 ± 0.09
CD-42 14707	129112145	0.99 ± 0.27	0.551264	0.551323	–	6649	–	0.61	–	44.16	–	−0.40	–
CD-48 13332	166463208	0.40 ± 0.04	0.579506	0.578464	6651 ± 349	6711	–	0.67	–	45.00	–	−1.79	–
GI Psc	406413012	0.42 ± 0.05	0.734199	0.736294	6487 ± 262	6477	–	0.71	–	59.77	–	−0.34	–

Notes: ID is the variable star identifier. TESS ID is the TESS Input Catalog identifier. Plx is the parallax in arcseconds. Period is the pulsation period in days (from the literature), and Period LS is the period determined from the Lomb–Scargle periodogram applied to TESS light curves. Columns with subscript 'lit' show values from the literature with their uncertainties when available. Columns with the subscript 'ANN' show values derived from our ANN.

While the temperature and luminosity predictions show reasonable agreement with literature values, mass estimates from the ANN are systematically lower by approximately 0.74 $M_\odot$. From stellar evolution, RR Lyrae stars are known to have masses in the range 0.5–0.8 $M_\odot$. In contrast, many literature values (from the TIC catalogue) report anomalously high masses (1.1–1.7 $M_\odot$), which is likely due to the use of isochrone or SED fitting methods that may not have accounted for the evolved, low-mass RR Lyrae stars.

The relatively large spread in metallicity ($\sigma = 0.83$ dex) and mean offset of 0.27 dex likely arises from degeneracies in light curve morphology, where variations in metallicity can mimic the effects of other parameters. Furthermore, literature metallicities are derived from diverse sources and methods, which introduces inconsistencies when used as a reference benchmark.

Overall, the ANN method demonstrates reasonable predictive power for effective temperature and luminosity, while systematic shifts in mass and scatter in metallicity reflect a combination of astrophysical degeneracies and uncertainties in the literature values. This supports our approach of deriving fundamental stellar parameters directly from time-series photometry while also highlighting the need for caution when comparing to literature sources that may not be tailored for RR Lyrae-type variables.

4.3. Period–Luminosity–Metallicity (PLZ) Relation

Using the stellar parameters inferred from the TESS RRab light curves via the trained ANN, we constructed a new empirical period–luminosity–metallicity (PLZ) relation. We fitted the logarithm of the luminosity as a linear function of the pulsation period and metallicity:

$$\log(L/L_\odot) = a\log(P/\text{days}) + b\,[\text{Fe/H}] + c, \tag{7}$$

where P is the pulsation period in days, and [Fe/H] is the derived metallicity from the ANN-inferred Z values. The coefficients a, b, and c were determined using a least-squares fitting.

$$a = 1.458 \pm 0.028, \qquad b = -0.068 \pm 0.007, \qquad c = 2.040 \pm 0.007. \tag{8}$$

Figure 6 presents a plot of the ANN predicted luminosities as a function of period and metallicity, which was overlaid with the fitted PLZ plane. The scatter of individual stars around the plane is shown as a color-coded scatter plot with the color scale representing [Fe/H]. Stars with lower metallicity (bluer colors) lie systematically higher in luminosity at fixed period, which is consistent with theoretical expectations.

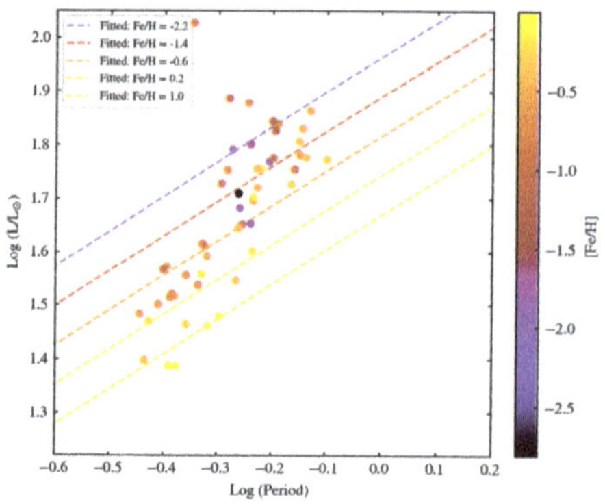

Figure 6. Period–luminosity–metallicity (PLZ) relation derived from ANN-inferred stellar parameters. The background shows the fitted plane, while the scatter points represent individual stars color-coded by metallicity [Fe/H].

5. Conclusions

In this study, we developed and validated an artificial neural network (ANN)-based framework to infer the physical parameters of fundamental-mode RR Lyrae (RRab) stars directly from their light curves. Using a synthetic grid of pulsation models, we trained a feedforward ANN to learn the inverse mapping from I band light-curve morphology to stellar parameters: mass (M), luminosity ($\log L$), effective temperature ($\log T_{\rm eff}$), and metallicity (Z). The network achieved high accuracy in recovering these parameters in synthetic self-inversion tests, demonstrating strong internal consistency and robustness across a wide parameter space.

The self-inversion test, conducted on a hold-out test set of 1,715 models not seen during training, showed good agreement between predicted and true values, particularly for $\log T_{\rm eff}$ and $\log L$. However, the performance was comparatively poorer for M and Z, which was likely due to degeneracies in the light-curve morphology, where multiple combinations of parameters can lead to similar pulsation features.

We then applied the trained parameter estimator model to observed RRab stars from the *TESS* mission. After preprocessing the light curves, including flux-to-magnitude conversion, phase folding, Fourier smoothing, extinction correction, and distance modulus calibration, we extracted stellar parameters from the light curves of individual stars. The values of M, $\log L$, $\log T_{\rm eff}$, and Z, inferred using the model, were used to compute surface gravity ($\log g$) and iron abundance [Fe/H], providing a complete physical characterization based solely on photometric data.

We compared the ANN-inferred values with those available in the literature (TIC v8.2 catalog). While the ANN-predicted values for $\log T_{\rm eff}$ and $\log L$ generally aligned well with the catalog values, the agreement was notably worse for mass and metallicity. A significant fraction of the TIC-inferred masses for RRab stars was found to lie in the range 1.1–1.7 $M_\odot$, which is inconsistent with established evolutionary models for RR Lyrae stars that predict a mass range of 0.5–0.8 $M_\odot$. These discrepancies suggest that catalog masses—often derived from isochrone or SED fitting—may not be reliable for evolved horizontal-branch stars like RR Lyraes. Consequently, the differences are more likely due to limitations in the literature data rather than overfitting or failure of the ANN model.

Furthermore, a period–luminosity–metallicity (PLZ) relation was derived using the ANN-predicted parameters. While the PLZ relation is physically plausible, its interpretation is naturally constrained by the accuracy of the inferred stellar properties.

This work demonstrates that an ANN-based inversion of RRab light curves offers an alternative to traditional model fitting and spectroscopic analysis. In the future, the method can be extended to include multi-band light curves, overtone RRc stars, and different classes of pulsating variable stars like Cepheids and BL Herculis-type variables. This work is the first step and can provide a fast way to estimate global physical parameters when trained with an extended parameter space of RRab theoretical models. Uncertainty-aware architectures and larger, better-constrained training sets will further improve the robustness and applicability of such data-driven frameworks in time-domain stellar astrophysics.

Author Contributions: Conceptualization, H.P.S. and N.K.; methodology, N.K., H.P.S., O.M. and S.J.; software, N.K.; validation, N.K.; formal analysis, N.K.; investigation, N.K.; resources, H.P.S. and S.J.; data curation, N.K.; writing—original draft preparation, N.K.; writing—review and editing, H.P.S., O.M., S.J., K.T., P.P. and A.B.; visualization, N.K.; supervision, H.P.S.; project administration, H.P.S. All authors have read and agreed to the published version of the manuscript.

Funding: This research was funded by the BRICS STI Framework Programme grant "Search and Follow-up Studies of Time-domain Astronomical Sources using Sky Surveys, BRICS Telescopes and Artificial Intelligence" (SAPTARISI). H.P.S. and S.J. were supported by the Department of Science & Technology (DST) through grant number DST/ICD/BRICS/Call-5/SAPTARISI/2023(G). O.M. was supported by the Ministry of Science and Higher Education of the Russian Federation, according to the research project 13.2251.21.0177 (075-15-2022-1228). K.T. was supported by the National Natural Science Foundation of China (NSFC) under grant No. 12261141689.

Institutional Review Board Statement: Not applicable.

Data Availability Statement: The data will be made available on request.

Acknowledgments: N.K. acknowledges the use of High Performance Cluster facility, Pegasus of IUCAA, Pune for providing the computational resources.

Conflicts of Interest: The authors declare no conflicts of interest.

Abbreviations

The following abbreviations are used in this manuscript:

ANN	artificial neural network
PLZ	period–luminosity–metallicity (relation)
TESS	Transiting Exoplanet Survey Satellite

Notes

1. The RRab interpolator for generating RRab light curves for a given combination of physical parameters is publicly available at https://ann-interpolator.web.app/ (accessed on 22 June 2025).
2. https://lightkurve.github.io/lightkurve/ (accessed on 22 June 2025).
3. https://archive.stsci.edu/files/live/sites/mast/files/home/missions-and-data/active-missions/tess/_documents/TESS_Instrument_Handbook_v0.1.pdf (accessed on 22 June 2025).
4. See note 3 above.

References

1. Savino, A.; Koch, A.; Prudil, Z.; Kunder, A.; Smolec, R. The age of the Milky Way inner stellar spheroid from RR Lyrae population synthesis. *Astron. Astrophys.* **2020**, *641*, A96. [CrossRef]
2. Longmore, A.J.; Fernley, J.A.; Jameson, R.F. RR Lyrae stars in globular clusters: Better distances from infrared measurements? *Mon. Not. R. Astron. Soc.* **1986**, *220*, 279–287.
3. Bono, G.; Caputo, F.; Castellani, V.; Marconi, M.; Storm, J. Theoretical insights into the RR Lyrae K-band period-luminosity relation. *Mon. Not. R. Astron. Soc.* **2001**, *326*, 1183–1190.
4. Catelan, M.; Pritzl, B.J.; Smith, H.A. The RR Lyrae Period-Luminosity Relation. I. Theoretical Calibration. *Astrophys. J. Suppl. Ser.* **2004**, *154*, 633–649.
5. Sollima, A.; Cacciari, C.; Valenti, E. The RR Lyrae period-K-luminosity relation for globular clusters: An observational approach. *Mon. Not. R. Astron. Soc.* **2006**, *372*, 1675–1680.
6. Muraveva, T.; Palmer, M.; Clementini, G.; Luri, X.; Cioni, M.R.L.; Moretti, M.I.; Marconi, M.; Ripepi, V.; Rubele, S. New Near-infrared Period-Luminosity-Metallicity Relations for RR Lyrae Stars and the Outlook for Gaia. *Astrophys. J.* **2015**, *807*, 127.
7. Bhardwaj, A.; Rejkuba, M.; de Grijs, R.; Yang, S.C.; Herczeg, G.J.; Marconi, M.; Singh, H.P.; Kanbur, S.; Ngeow, C.C. RR Lyrae Variables in Messier 53: Near-infrared Period-Luminosity Relations and the Calibration Using Gaia Early Data Release 3. *Astron. J.* **2021**, *909*, 200.
8. Beaton, R.L.; Freedman, W.L.; Madore, B.F.; Bono, G.; Carlson, E.K.; Clementini, G.; Durbin, M.J.; Garofalo, A.; Hatt, D.; Jang, I.S.; et al. The Carnegie-Chicago Hubble Program. I. An Independent Approach to the Extragalactic Distance Scale Using Only Population II Distance Indicators. *Astrophys. J.* **2016**, *832*, 210.
9. Bhardwaj, A. High-precision distance measurements with classical pulsating stars. *J. Astrophys. Astron.* **2020**, *41*, 23.
10. Catelan, M. Horizontal branch stars: The interplay between observations and theory, and insights into the formation of the Galaxy. *Astrophys. Space Sci.* **2009**, *320*, 261–309.
11. Kunder, A.; Valenti, E.; Dall'Ora, M.; Pietrukowicz, P.; Sneden, C.; Bono, G.; Braga, V.F.; Ferraro, I.; Fiorentino, G.; Iannicola, G.; et al. Impact of Distance Determinations on Galactic Structure. II. Old Tracers. *Space Sci. Rev.* **2018**, *214*, 90.

12. Cacciari, C.; Corwin, T.M.; Carney, B.W. A Multicolor and Fourier Study of RR Lyrae Variables in the Globular Cluster NGC 5272 (M3). *Astron. J.* **2005**, *129*, 267–302.
13. Deb, S.; Singh, H.P. Physical parameters of the Small Magellanic Cloud RR Lyrae stars and the distance scale. *Mon. Not. R. Astron. Soc.* **2010**, *402*, 691–704.
14. Nemec, J.M.; Smolec, R.; Benkő, J.M.; Moskalik, P.; Kolenberg, K.; Szabó, R.; Kurtz, D.W.; Bryson, S.; Guggenberger, E.; Chadid, M.; et al. Fourier analysis of non-Blazhko ab-type RR Lyrae stars observed with the Kepler space telescope. *Mon. Not. R. Astron. Soc.* **2011**, *417*, 1022–1053.
15. Marconi, M.; Coppola, G.; Bono, G.; Braga, V.; Pietrinferni, A.; Buonanno, R.; Castellani, M.; Musella, I.; Ripepi, V.; Stellingwerf, R.F. On a New Theoretical Framework for RR Lyrae Stars. I. The Metallicity Dependence. *Astrophys. J.* **2015**, *808*, 50.
16. De Somma, G.; Marconi, M.; Molinaro, R.; Cignoni, M.; Musella, I.; Ripepi, V. An Extended Theoretical Scenario for Classical Cepheids. I. Modeling Galactic Cepheids in the Gaia Photometric System. *Astron. J. Suppl. Ser.* **2020**, *247*, 30.
17. De Somma, G.; Marconi, M.; Molinaro, R.; Ripepi, V.; Leccia, S.; Musella, I. An Updated Metal-dependent Theoretical Scenario for Classical Cepheids. *Astrophys. J. Suppl. Ser.* **2022**, *262*, 25.
18. Smolec, R.; Moskalik, P. Convective Hydrocodes for Radial Stellar Pulsation. Physical and Numerical Formulation. *Acta Astron.* **2008**, *58*, 193–232.
19. Paxton, B.; Bildsten, L.; Dotter, A.; Herwig, F.; Lesaffre, P.; Timmes, F. Modules for Experiments in Stellar Astrophysics (MESA). *Astrophys. J. Suppl. Ser.* **2011**, *192*, 3.
20. Paxton, B.; Cantiello, M.; Arras, P.; Bildsten, L.; Brown, E.F.; Dotter, A.; Mankovich, C.; Montgomery, M.H.; Stello, D.; Timmes, F.X.; et al. Modules for Experiments in Stellar Astrophysics (MESA): Planets, Oscillations, Rotation, and Massive Stars. *Astrophys. J. Suppl. Ser.* **2013**, *208*, 4.
21. Paxton, B.; Marchant, P.; Schwab, J.; Bauer, E.B.; Bildsten, L.; Cantiello, M.; Dessart, L.; Farmer, R.; Hu, H.; Langer, N.; et al. Modules for Experiments in Stellar Astrophysics (MESA): Binaries, Pulsations, and Explosions. *Astrophys. J. Suppl. Ser.* **2015**, *220*, 15.
22. Paxton, B.; Schwab, J.; Bauer, E.B.; Bildsten, L.; Blinnikov, S.; Duffell, P.; Farmer, R.; Goldberg, J.A.; Marchant, P.; Sorokina, E.; et al. Modules for Experiments in Stellar Astrophysics (MESA): Convective Boundaries, Element Diffusion, and Massive Star Explosions. *Astrophys. J. Suppl. Ser.* **2018**, *234*, 34.
23. Paxton, B.; Smolec, R.; Schwab, J.; Gautschy, A.; Bildsten, L.; Cantiello, M.; Dotter, A.; Farmer, R.; Goldberg, J.A.; Jermyn, A.S.; et al. Modules for Experiments in Stellar Astrophysics (MESA): Pulsating Variable Stars, Rotation, Convective Boundaries, and Energy Conservation. *Astrophys. J. Suppl. Ser.* **2019**, *243*, 10.
24. Torres, G.; Andersen, J.; Giménez, A. Accurate masses and radii of normal stars: Modern results and applications. *Astron. Astrophys. Rev.* **2010**, *18*, 67–126.
25. Southworth, J. DEBCat: A Catalog of Detached Eclipsing Binary Stars. In Proceedings of the Living Together: Planets, Host Stars and Binaries, Litomyšl, Czech Republic, 8–12 September 2014; Rucinski, S.M., Torres, G., Zejda, M., Eds.; Astronomical Society of the Pacific Conference Series; Astronomical Society of the Pacific: San Francisco, CA, USA, 2015; Volume 496, p. 164.
26. Grison, P.; Beaulieu, J.P.; Pritchard, J.D.; Tobin, W.; Ferlet, R.; Vidal-Madjar, A.; Guibert, J.; Alard, C.; Moreau, O.; Tajahmady, F.; et al. EROS catalogue of eclipsing binary stars in the bar of the Large Magellanic Cloud. *Astron. Astrophys. Suppl.* **1995**, *109*, 447–469.
27. Alcock, C.; Allsman, R.A.; Alves, D.R.; Axelrod, T.S.; Becker, A.C.; Basu, A.; Baskett, L.; Bennett, D.P.; Cook, K.H.; Freeman, K.C.; et al. The RR Lyrae Population of the Galactic Bulge from the MACHO Database: Mean Colors and Magnitudes. *Astrophys. J.* **1998**, *492*, 190–199. [CrossRef]
28. Rucinski, S.M.; Maceroni, C. Eclipsing Binaries in the OGLE Variable Star Catalogs. V. Long-Period EB-Type Light Curve Systems in the Small Magellanic Cloud and the PLC-βRelation. *Astron. J.* **2001**, *121*, 254–266.
29. Pojmanski, G. The All Sky Automated Survey. Catalog of Variable Stars. I. 0 h–6 hQuarter of the Southern Hemisphere. *Acta Astron.* **2002**, *52*, 397–427.
30. Alonso, R.; Brown, T.M.; Torres, G.; Latham, D.W.; Sozzetti, A.; Mandushev, G.; Belmonte, J.A.; Charbonneau, D.; Deeg, H.J.; Dunham, E.W.; et al. TrES-1: The Transiting Planet of a Bright K0 V Star. *Astrophys. J. Lett.* **2004**, *613*, L153–L156.
31. Bakos, G.; Noyes, R.W.; Kovács, G.; Stanek, K.Z.; Sasselov, D.D.; Domsa, I. Wide-Field Millimagnitude Photometry with the HAT: A Tool for Extrasolar Planet Detection. *Publ. Astron. Soc. Pac.* **2004**, *116*, 266–277.
32. Loeillet, B.; Bouchy, F.; Deleuil, M.; Royer, F.; Bouret, J.C.; Moutou, C.; Barge, P.; de Laverny, P.; Pont, F.; Recio-Blanco, A.; et al. Doppler search for exoplanet candidates and binary stars in a CoRoT field using a multi-fiber spectrograph. I. Global analysis and first results. *Astron. Astrophys.* **2008**, *479*, 865–875.
33. Matijevič, G.; Prša, A.; Orosz, J.A.; Welsh, W.F.; Bloemen, S.; Barclay, T. Kepler Eclipsing Binary Stars. III. Classification of Kepler Eclipsing Binary Light Curves with Locally Linear Embedding. *Astron. J.* **2012**, *143*, 123.
34. Das, S.; Bhardwaj, A.; Kanbur, S.M.; Singh, H.P.; Marconi, M. On the variation of light-curve parameters of RR Lyrae variables at multiple wavelengths. *Mon. Not. R. Astron. Soc.* **2018**, *481*, 2000–2017.

35. Kumar, N.; Prugniel, P.; Singh, H.P. Physical parameters of stars in NGC 6397 using ANN-based interpolation and full spectrum fitting. *New Astron.* **2025**, *119*, 102416. [CrossRef]
36. Bellinger, E.P.; Angelou, G.C.; Hekker, S.; Basu, S.; Ball, W.H.; Guggenberger, E. Fundamental Parameters of Main-Sequence Stars in an Instant with Machine Learning. *Astrophys. J.* **2016**, *830*, 31.
37. Kumar, N.; Bhardwaj, A.; Singh, H.P.; Das, S.; Marconi, M.; Kanbur, S.M.; Prugniel, P. Predicting light curves of RR Lyrae variables using artificial neural network based interpolation of a grid of pulsation models. *Mon. Not. R. Astron. Soc.* **2023**, *522*, 1504–1520.
38. Bellinger, E.P.; Kanbur, S.M.; Bhardwaj, A.; Marconi, M. When a period is not a full stop: Light-curve structure reveals fundamental parameters of Cepheid and RR Lyrae stars. *Mon. Not. R. Astron. Soc.* **2020**, *491*, 4752–4767.
39. Kumar, N.; Bhardwaj, A.; Singh, H.P.; Rejkuba, M.; Marconi, M.; Prugniel, P. Multiwavelength photometric study of RR lyrae variables in the globular cluster NGC 5272 (Messier 3). *Mon. Not. R. Astron. Soc.* **2024**, *531*, 2976–2997.
40. van Albada, T.S.; Baker, N. On the Masses, Luminosities, and Compositions of Horizontal-Branch Stars. *Astrophys. J.* **1971**, *169*, 311.
41. Ricker, G.R.; Winn, J.N.; Vanderspek, R.; Latham, D.W.; Bakos, G.A.; Bean, J.L.; Berta-Thompson, Z.K.; Brown, T.M.; Buchhave, L.; Butler, N.R.; et al. Transiting Exoplanet Survey Satellite (TESS). *J. Astron. Telesc. Instrum. Syst.* **2015**, *1*, 014003. [CrossRef]
42. Molnár, L.; Bódi, A.; Pál, A.; Bhardwaj, A.; Hambsch, F.; Benkő, J.M.; Derekas, A.; Ebadi, M.; Joyce, M.; Hasanzadeh, A.; et al. First Results on RR Lyrae Stars with the TESS Space Telescope: Untangling the Connections between Mode Content, Colors, and Distances. *Astrophys. J. Suppl. Ser.* **2021**, *258*, 8. [CrossRef]
43. Stassun, K.G.; Oelkers, R.J.; Pepper, J.; Paegert, M.; De Lee, N.; Torres, G.; Latham, D.W.; Charpinet, S.; Dressing, C.D.; Huber, D.; et al. The TESS Input Catalog and Candidate Target List. *Astron. J.* **2018**, *156*, 102.
44. Paegert, M.; Stassun, K.G.; Collins, K.A.; Pepper, J.; Torres, G.; Jenkins, J.; Twicken, J.D.; Latham, D.W. TESS Input Catalog versions 8.1 and 8.2: Phantoms in the 8.0 Catalog and How to Handle Them. *arXiv* **2021**, arXiv:2108.04778.
45. Stassun, K.G.; Oelkers, R.J.; Paegert, M.; Torres, G.; Pepper, J.; De Lee, N.; Collins, K.; Latham, D.W.; Muirhead, P.S.; Chittidi, J.; et al. The Revised TESS Input Catalog and Candidate Target List. *Astron. J.* **2019**, *158*, 138.
46. Lightkurve Collaboration; Cardoso, J.V.d.M.; Hedges, C.; Gully-Santiago, M.; Saunders, N.; Cody, A.M.; Barclay, T.; Hall, O.; Sagear, S.; Turtelboom, E.; et al. *Lightkurve: Kepler and TESS Time Series Analysis in Python*; Astrophysics Source Code Library. 2018. Available online: https://ui.adsabs.harvard.edu/abs/2018ascl.soft12013L/abstract (accessed on 15 June 2025).
47. Cybenko, G. Approximation by superpositions of a sigmoidal function. *Math. Control Signals Syst.* **1989**, *2*, 303–314. [CrossRef]
48. Hornik, K.; Stinchcombe, M.; White, H. Multilayer feedforward networks are universal approximators. *Neural Netw.* **1989**, *2*, 359–366. [CrossRef]
49. Hornik, K. Approximation capabilities of multilayer feedforward networks. *Neural Netw.* **1991**, *4*, 251–257. [CrossRef]
50. Kingma, D.P.; Ba, J. Adam: A Method for Stochastic Optimization. *arXiv* **2014**, arXiv:1412.6980.
51. Jenkins, J.M.; Twicken, J.D.; McCauliff, S.; Campbell, J.; Sanderfer, D.; Lung, D.; Mansouri-Samani, M.; Girouard, F.; Tenenbaum, P.; Klaus, T.; et al. The TESS science processing operations center. In Proceedings of the Software and Cyberinfrastructure for Astronomy IV, Edinburgh, UK, 17–21 July 2016; Chiozzi, G., Guzman, J.C., Eds.; Society of Photo-Optical Instrumentation Engineers (SPIE) Conference Series; SPIE: Bellingham, WA, USA, 2016; Volume 9913, p. 99133E. [CrossRef]
52. Lomb, N.R. Least-squares frequency analysis of unequally spaced data. *Astrophys. Space Sci.* **1976**, *39*, 447–462. [CrossRef]
53. Scargle, J.D. Studies in astronomical time series analysis. II. Statistical aspects of spectral analysis of unevenly spaced data. *Astrophys. J.* **1982**, *263*, 835–853.
54. Schlafly, E.F.; Finkbeiner, D.P. Measuring Reddening with Sloan Digital Sky Survey Stellar Spectra and Recalibrating SFD. *Astrophys. J.* **2011**, *737*, 103.
55. Piersanti, L.; Straniero, O.; Cristallo, S. A method to derive the absolute composition of the Sun, the solar system, and the stars*. *Astron. Astrophys.* **2007**, *462*, 1051–1062. [CrossRef]
56. Asplund, M.; Grevesse, N.; Sauval, A.J. The Solar Chemical Composition. *ASPC Conf.* **2005**, *336*, 25. [CrossRef]
57. Marconi, M.; Minniti, D. Gauging the Helium Abundance of the Galactic Bulge RR Lyrae Stars. *Astrophys. J. Lett.* **2018**, *853*, L20.

Disclaimer/Publisher's Note: The statements, opinions and data contained in all publications are solely those of the individual author(s) and contributor(s) and not of MDPI and/or the editor(s). MDPI and/or the editor(s) disclaim responsibility for any injury to people or property resulting from any ideas, methods, instructions or products referred to in the content.

universe

MDPI

Article

Spectroscopic Observations and Emission-Line Diagnoses for H II Regions in the Late-Type Spiral Galaxy NGC 2403

Qi-Ming Wu [1,2], Ye-Wei Mao [1,2,*], Lin Lin [3], Hu Zou [4] and Shu-Ting Wang [1,2]

1 Center for Astrophysics, Guangzhou University, Guangzhou 510006, China; qimingwu@e.gzhu.edu.cn (Q.-M.W.); wangshuting@e.gzhu.edu.cn (S.-T.W.)
2 Department of Astronomy, School of Physics and Materials Science, Guangzhou University, Guangzhou 510006, China
3 Shanghai Astronomical Observatory, Chinese Academy of Sciences, Shanghai 200030, China; linlin@shao.ac.cn
4 National Astronomical Observatories, Chinese Academy of Sciences, Beijing 100101, China; zouhu@nao.cas.cn
* Correspondence: ywmao@gzhu.edu.cn

Abstract: Being ionized nebulae where star formation events take place, H II regions are not only natural laboratories for studying physical processes of star formation and photoionization but also signatures reflecting evolution of their internal stellar populations and hosting galaxies. In this paper, we present a comprehensive analysis of spectral emission-line data for H II regions in the nearby spiral galaxy NGC 2403, aimed at gaining deep insight into underlying properties and evolution for the H II regions and the galaxy. The spectroscopic data are obtained through observations with the 2.16 m telescope at National Astronomical Observatories of China and a collection of published data in the literature. Photoionization modeling is combined in the analysis for diagnosing the spectral features and interpreting the observational data with certain physical mechanisms. Results of this work not only involve estimates of a set of parameters such as metallicity, the ionization parameter, etc., and evolution stages for the H II regions in NGC 2403 but also reveal distinct characteristics of different spectral features and their sensitivities to specific parameters, which provides an instructive implication for proper usages of emission-line diagnostics for H II regions or galaxies nearby and far away.

Keywords: H II regions; ISM: lines and bands; galaxies: abundances; galaxies: individual (NGC 2403); galaxies: ISM

1. Introduction

Being gaseous nebulae irradiated by recently formed massive stars, H II regions are conspicuous sites where star formation activities accompanying photoionization processes occur, and young stellar populations surrounded by ionized interstellar medium reside. Spectra of H II regions contain a great number of emission lines, which serve as a useful tool for probing not only astrophysical mechanisms/properties in H II regions but also evolution history of host galaxies (Peimbert et al. [1], Maiolino and Mannucci [2], Kewley et al. [3]).

Metallicity is an important parameter derivable from emission-line features for H II regions. It behaves like a signature of element enrichment with star formation, and hence its spatial distributions are an imprint of chemical evolution for galaxies. The most accurate method for deriving metallicity from spectral emission lines is based on estimates of electron temperature (T_e) and referred to as the T_e or direct method (Peimbert et al. [1]). However, the T_e method requires measurements of auroral lines such as [O III]λ4363,

https://doi.org/10.3390/universe11080280

[N II]λ5755, or [S III]λ6312, which are intrinsically weak in spectra and even undetectable at high metallicity. In this case, strong collisionally excited emission lines such as [O II]λ3727, [O III]$\lambda\lambda$4959, 5007, [N II]$\lambda\lambda$6548, 6583, and [S II]$\lambda\lambda$6717, 6731 are applied to estimating metallicity with empirical or theoretical calibrations, and they have provided more widely used diagnostics in studies of external galaxies (Alloin et al. [4], Pagel et al. [5], Pettini and Pagel [6], Pilyugin and Thuan [7], Bresolin [8]).

In addition to metallicity, the collisionally excited emission lines have been found to also relate with the ionization parameter and electron density at different levels of sensitivities. Some spectral indices combining certain emission lines have been adopted to probe the physical parameters for H II regions or emission-line galaxies (e.g., [O III]λ5007/[O II]λ3727 or [S III]$\lambda\lambda$9069, 9531/[S II]$\lambda\lambda$6717, 6731 for probing the ionization parameter, Kewley and Dopita [9], Morisset et al. [10]; [O II]λ3726/[O II]λ3729 or [S II]λ6717/[S II]λ6731 for probing electron density, Wang et al. [11]).[1]

Since H II regions are excited by young massive stars, aging of the ionizing sources is considered to have an effect on the excitation and the subsequent spectral features (Levesque et al. [12], Xiao et al. [13]). Variations of commonly adopted diagnostic indices with stellar population age can be non-negligible, as they may affect the inferred metallicity, ionization parameter, electron density, and other physical properties. Nevertheless, the impact of age on the emission lines of H IIregions has yet to be systematically addressed in observational studies of galaxies.

Kong et al. [14] conducted a spectroscopic survey of 20 nearby galaxies and obtained a substantial collection of H II region spectra. These data provide critical information on metallicity, ionization parameters, and dust extinction, offering observational constraints for theoretical models of galaxy formation, chemical evolution, and photoionization. As one of the target galaxies in this project, NGC 2403 is a late-type spiral galaxy (classified into the SAB(s)cd morphological type) located in the M81 galaxy group at the distance ~3.18 Mpc from the Milky Way (Tully et al. [15]). Basic information on astrometry for this galaxy includes the celestial coordinates R.A. = 07:36:51.3, Dec. = +65:36:09.7 (Skrutskie et al. [16]), the apparent size $R_{25} = \sim 10'.93$, the inclination $\sim 60^\circ$, and the position angle $\sim 126^\circ$ (de Vaucouleurs et al. [17]). There are a number of bright H II regions widespread in NGC 2403, making it possible to conduct a spatially resolved spectroscopic analysis. Mao et al. [18] performed a spatially resolved study of 11 large, bright H II regions in NGC 2403 using the Kong et al. [14] dataset. They examined the performance of various metallicity calibration indicators within individual H II regions.

In the present work, we target H II regions in the late-type spiral galaxy NGC 2403 and carry out an observational investigation in combination with modeling. Spectroscopy is taken for the sampled H II regions and aimed at a comprehensive inspection of emission-line characteristics related to underlying physical properties. In particular, we attempt to derive the parameters from certain diagnostics on the one hand and scrutinize specific influences of certain parameters on different emission-line features on the other hand.

Subsequent content of this paper is structured as follows. Section 2 presents observations, data reduction, measurements, and ancillary data from the literature. Section 3 introduces the photoionization model employed in this study. Results are described in Section 4, followed by a discussion about implications and suggestions in Section 5. Section 6 in the end summarizes conclusions drawn in this work.

2. Data

2.1. Observations

A total of 84 H II regions in NGC 2403 were selected from the catalog of Sivan et al. [19] and observed with the 2.16 m telescope at the Xinglong Station of National Astronomical Observatories of China (NAOC) as part of a spectroscopic survey of H II regions in 20 nearby large galaxies (Kong et al. [14]). The 2.16 m telescope equips the Optomechanics Research Inc. (OMR) long-slit spectrograph with a TEKTRONIX TEK1024, AR-coated back-illuminated CCD. A 200 Å/mm dispersion grating, blazed at ∼5500 Å, was used in the spectrograph with a 4′-length slit, providing a dispersion sampling of 4.8 Å per pixel and a spectral resolution of 500–550 in $\lambda/\Delta\lambda$ at 5000 Å with a wavelength coverage of 3500–8000 Å (Fan et al. [20]).

The long-slit spectroscopic observations were carried out over several nights from January 2007 to December 2008 (Kong et al. [14], Mao et al. [18]). During the observing runs, the slit width was set to 2″.5 to match the average local seeing disk. The position angle of the long slit was oriented to cross as many H II regions as possible at one exposure frame. The exposure time for each slit position was 1800 s × 2 or 1200 s × 3 to achieve a suitable signal-to-noise (S/N) ratio. Instrument bias and dome flats were recorded at the beginning and the end of each night. A He-Ar lamp was observed immediately after observing each object at the same position for wavelength calibration. Spectrophotometric standard stars including Feige 34 and He 3 (Corbin and Warren [21]) were observed in each night for flux calibration. Table 1 lists statistics of the observations. The orientations of the spectrographic slit and the locations of the H II regions are shown in Figure 1.

Table 1. Observations.

Slit ID. (1)	Obs. Date (2)	Slit Angle (3)	R.A. (4)	Decl. (5)	Exp. Time (6)	Airmass (7)
NGC 2403-A	16 January 2007	−33.7	07:37:00	+65:36:08	1800 × 2	1.1128
NGC 2403-C	17 January 2007	+40.4	07:36:52	+65:36:11	1800 × 2	1.1056
NGC 2403-E	18 January 2007	−52.0	07:37:07	+65:36:39	1800 × 2	1.1059
NGC 2403-G	18 January 2007	−42.5	07:36:42	+65:36:02	1800 × 2	1.1067
NGC 2403-H	01 January 2008	+9.7	07:36:21	+65:38:01	1800 × 2	1.1064
NGC 2403-I	01 January 2008	+9.7	07:37:17	+65:32:42	1800 × 2	1.1588
NGC 2403-O	02 January 2008	+68.4	07:36:32	+65:39:09	1800 × 2	1.1218
NGC 2403-L	03 January 2008	−80.2	07:37:03	+65:34:02	1800 × 2	1.1158
NGC 2403-K	04 January 2008	−83.9	07:36:50	+65:35:01	1800 × 2	1.1118
NGC 2403-N	04 January 2008	−77.0	07:36:48	+65:33:25	1800 × 2	1.1422
NGC 2403-Q	05 January 2008	+74.5	07:36:41	+65:39:06	1800 × 2	1.1127
NGC 2403-S	06 January 2008	−26.1	07:36:23	+65:36:19	1800 × 2	1.1055
NGC 2403-U	07 January 2008	−71.0	07:36:51	+65:36:45	1800 × 2	1.1218
NGC 2403-01	29 November 2008	−70.0	07:36:51	+65:36:09	1200 × 3	1.1484
NGC 2403-04	29 November 2008	+10.0	07:36:51	+65:36:09	1200 × 3	1.1052
NGC 2403-05	29 November 2008	+13.0	07:36:51	+65:36:09	1200 × 3	1.1524
NGC 2403-02	30 November 2008	+90.8	07:36:51	+65:36:09	1200 × 3	1.1122
NGC 2403-03	01 December 2008	+63.0	07:37:09	+65:38:02	1200 × 3	1.1124

Notes. Columns: (1) the identifiers of the slit coverages; (2) the observation period; (3) the rotate angle; (4)–(5) R.A., Decl. for the central position of the slit; (6)–(7) the exposure time for each observation and air mass.

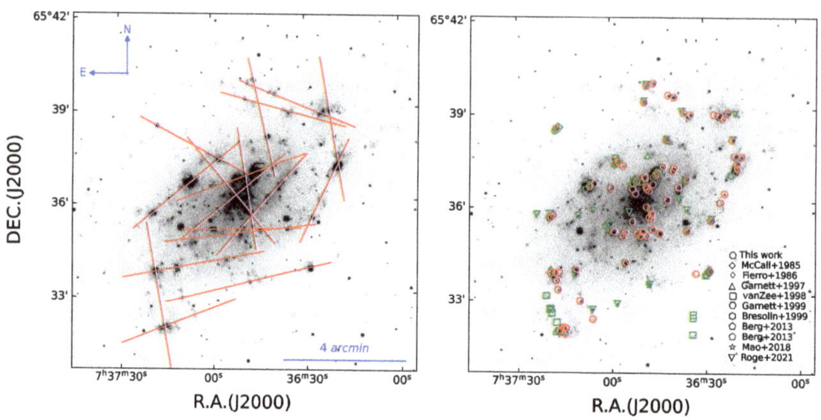

Figure 1. (**Left panel**): Hα narrow-band image for NGC 2403 taken with the 2.1 m telescope at Kitt Peak National Observatory, with the coverages of the spectrographic slit (red solid lines) superimposed. The information of each of the slit coverages are listed in Table 1. (**Right panel**): The same image as the left panel but with H II regions studied in this work marked by different symbols in different colors; the H II regions observed with the 2.16 m telescope at NAOC are marked with red open circles; those collected from other studies are colored in green and symbolized as listed in the lower-right corner of the panel. The scale in the bottom right corner in the left panel indicates the 4 arcmin length which is the same in the right panel. In each of the panels, north is up, and east is to the left.

2.2. Data Reduction

Data reduction was conducted using PyRAF 2.2.0, a Python-based interface to IRAF V2.17,[2] employing the CCDRED, TWODSPEC, and ONEDSPEC packages. The conventional routine including bias removal, flat-field correction, and cosmic-ray rejection was implemented in the two-dimensional (2D) data in the form of a space–dispersion plane. One-dimensional spectra were extracted using $10''$ apertures, corresponding to the $\sim$154 pc local scale (by taking the distance $\sim$3.18 Mpc addressed in Tully et al. [15] into account). These apertures were placed along the spatial axis in the spatial–dispersion plane of the 2D spectra to extract the H II region profiles. When placing the apertures, we prioritized centering them on the peaks of the Hα emission, provided that adjacent apertures would not overlap. In cases where two nearby Hα peaks were closely spaced, we slightly shifted the aperture centers to avoid overlap, while still ensuring coverage of the emission peaks.

Spectral trajectories along the dispersion were traced by sampling and fitting continuum points with certain tracing function; the continuum points were obtained by summing enough dispersion lines; for some H II regions with very weak continua, spectra for the standard stars were used as reference for tracing the dispersion trajectories. Background levels were estimated in blank areas along the spatial axis of the 2D data and subtracted from the spectra of the objects during the spectral extraction process.

For each of the extracted spectra, the dispersion scale was calibrated to wavelength by referencing the emission-line spectra of the He-Ar lamp. Flux calibration was performed by adopting the spectra of the spectrophotometric standard stars and the atmospheric extinction curve at the XingLong station. During our observations, the air mass ranged from 1.105 to 1.159 (mean 1.120). A fixed $2''.5$ slit width was adopted to match the average seeing of $2''.0$ (Fan et al. [20]). The slit was not oriented at the parallactic angle in order to maximize target coverage. Following Filippenko [22], we estimated differential atmospheric refraction (DAR) offsets at $\sec z \approx 1.15$ to be $+0''.53$ at $[OII]\lambda3727$ and $-0''.27$ at $[SII]\lambda6731$, relative to 5000 Å. At $\sec z \approx 1.10$, the offsets were $+0''.43$ and $-0''.22$, respectively. Because the slit width direction was not always perpendicular to the atmospheric dispersion direction, the actual centroid shifts were smaller than the full DAR. At 3721 Å, all DAR-induced shifts were substantially smaller than both the local seeing and the slit width. Consequently, no DAR or slit-loss corrections were applied in this work. All of the calibrated spectra were corrected for galactic foreground extinction, by utilizing the Cardelli et al. [23] (hereafter denoted as CCM) extinction curve and the total-to-selective extinction ratio $R_V = 3.1$. The color excess of the galactic extinction $E(B-V)_{GAL} = 0.04$ mag for NGC 2403 was obtained from the Schlegel et al. [24] galactic dust map and applied to the correction.

2.3. Emission-Line Measurements

The STARLIGHT software (Cid Fernandes et al. [25]) was employed to reproduce the underlying stellar continuum, for the purpose of subtraction the continuum from each of the calibrated spectra to obtain sole emission-line spectra. Before the continuum fits, all of the observed spectra were corrected for redshift (shifted to the rest frame) and resampled to 1 Å per pixel using the SpectRes code (Carnall [26]), suiting the requirement of STARLIGHT. The model was composed of simple stellar populations (SSPs) selected from the Bruzual and Charlot [27] spectral library of stellar population synthesis, encompassing 75 ages spanning from 1 Myr to 18 Gyr and 3 metallicities (Z = 0.008, 0.02, 0.05), on the basis of the Chabrier [28] initial mass function (IMF) and the Padova 1994 evolutionary tracks (Le Borgne et al. [29]). Dust attenuation was imposed on the model spectra with the CCM extinction law taken into account. The fits for the observed spectra were performed through a χ^2-minimization approach using the Markov Chain Monte Carlo (MCMC) algorithm, and the best-fit model spectra were obtained.

Emission-line fluxes were measured in the continuum-subtracted spectra via Gaussian-profile fits. Each of the emission lines [O II]λ3727, Hβ, and [O III]λ5007 was fitted with a single Gaussian profile; the three blended lines [N II]$\lambda\lambda$6548, 6583 + Hα as well as the two blended lines [S II]$\lambda\lambda$6717, 6731 were deblended by fitting them with multiple Gaussian components. The flux ratio of [O III]λ4959 to [O III]λ5007 was fixed to 1:3, and so was [N II]λ6548 to [N II]λ6583.

Uncertainties of the emission-line fluxes were estimated using the expression in Mao et al. [18], Gonzalez-Delgado et al. [30]:

$$\sigma_{\text{line}} = \sigma_{\text{cont}} N_{\text{pix}}^{1/2} [1 + EW/(N\Delta_\lambda)]^{1/2}, \tag{1}$$

where σ_{cont} is the standard deviation of the continuum close to the emission line, N_{pix} is the number of pixels covering the measured emission line, EW is the equivalent width of the emission line, and Δ_λ is the dispersion in units of Å/pixel.

The measured emission-line fluxes were corrected for dust attenuation using the Balmer decrement, following the prescription in Calzetti [31]:

$$\frac{(L_{\text{H}\alpha}/L_{\text{H}\beta})_{obs}}{(L_{\text{H}\alpha}/L_{\text{H}\beta})_{ini}} = 10^{0.4E(B-V)[k(\text{H}\beta)-k(\text{H}\alpha)]}, \tag{2}$$

where $(L_{\text{H}\alpha}/L_{\text{H}\beta})_{obs}$ is the observed Hα-to-Hβ flux ratio, $(L_{\text{H}\alpha}/L_{\text{H}\beta})_{ini} = 2.86$ is the assumed intrinsic Hα-to-Hβ flux ratio ratio in accordance with the Case B recombination at the electron temperature $T_e = 10^4$ K and the electron density $n_e = 100$ cm^{-3} (Osterbrock and Ferland [32]), $k(\text{H}\beta) - k(\text{H}\alpha) = 1.074$ and $R_V = 3.1$ are quoted from the CCM extinction law.

The attenuation-corrected fluxes for the emission lines are listed in Table 2.

Table 2. Attenuation-corrected emission-line fluxes.

ID (1)	R.A. (J2000) (2)	Dec. (J2000) (3)	R/R_{25} (4)	Hα (5)	[O II]λ3727 (6)	[O III]λ5007 (7)	[N II]λ6583 (8)	[S II]λλ6717,31 (9)	A_V (mag) (10)	Hβ 10^{-15}(erg s^{-1} cm^{-2}) (11)
1	7:37:05.938	+65:32:20.727	0.512	2.860 ± 0.804	–	0.662 ± 0.213	0.288 ± 0.087	0.644 ± 0.175	0.559 ± 0.208	6.1 ± 1.4
2	7:37:14.724	+65:32:03.740	0.529	2.727 ± 0.018	2.699 ± 0.073	2.704 ± 0.019	0.265 ± 0.003	0.581 ± 0.010	–	35.4 ± 0.2
3	7:37:15.797	+65:32:01.689	0.532	2.843 ± 0.031	3.136 ± 0.143	1.493 ± 0.018	0.338 ± 0.007	0.723 ± 0.012	–	18.9 ± 0.2
4	7:37:16.973	+65:31:58.414	0.538	2.860 ± 0.123	2.413 ± 0.167	2.575 ± 0.125	0.292 ± 0.013	0.427 ± 0.019	0.385 ± 0.032	55.2 ± 2.0
5	7:37:17.435	+65:38:30.523	0.642	2.860 ± 0.113	1.840 ± 0.107	4.109 ± 0.184	0.098 ± 0.005	0.161 ± 0.011	0.239 ± 0.029	74.0 ± 2.4
6	7:36:57.819	+65:37:24.469	0.257	2.860 ± 0.070	3.503 ± 0.122	1.306 ± 0.036	0.467 ± 0.012	0.381 ± 0.013	0.750 ± 0.018	144.0 ± 2.9
7	7:36:55.760	+65:37:15.956	0.214	2.725 ± 0.083	–	1.501 ± 0.046	0.710 ± 0.030	0.902 ± 0.038	–	5.8 ± 0.2
8	7:36:35.620	+65:35:08.406	0.332	2.860 ± 0.321	3.271 ± 0.518	1.506 ± 0.192	0.482 ± 0.055	0.689 ± 0.073	0.095 ± 0.083	43.2 ± 4.0
9	7:36:44.249	+65:35:04.334	0.240	2.860 ± 0.226	3.025 ± 0.530	0.866 ± 0.078	0.630 ± 0.051	0.730 ± 0.057	0.588 ± 0.059	48.8 ± 3.2
10	7:36:49.959	+65:35:07.739	0.174	2.860 ± 0.368	1.837 ± 0.468	0.850 ± 0.124	0.640 ± 0.083	0.821 ± 0.099	0.273 ± 0.095	27.2 ± 2.9
11	7:36:51.597	+65:35:06.539	0.163	2.860 ± 0.762	–	0.523 ± 0.160	0.580 ± 0.156	0.810 ± 0.201	0.436 ± 0.198	18.3 ± 4.0
12	7:36:59.240	+65:35:09.214	0.130	2.860 ± 0.354	3.903 ± 0.682	0.609 ± 0.086	0.683 ± 0.085	0.812 ± 0.094	0.806 ± 0.092	51.8 ± 5.3
13	7:36:59.389	+65:35:09.230	0.130	2.860 ± 0.298	2.038 ± 0.386	0.645 ± 0.076	0.533 ± 0.057	0.520 ± 0.051	0.216 ± 0.077	35.9 ± 3.1
14	7:36:49.283	+65:36:52.725	0.100	2.860 ± 0.234	2.106 ± 0.260	0.771 ± 0.071	0.642 ± 0.053	0.722 ± 0.055	0.513 ± 0.061	36.6 ± 2.5
15	7:36:48.835	+65:36:35.144	0.055	2.860 ± 0.215	1.952 ± 0.212	0.712 ± 0.062	0.574 ± 0.044	0.621 ± 0.045	0.039 ± 0.056	15.8 ± 1.0
16	7:36:48.686	+65:36:30.517	0.045	2.540 ± 0.050	1.957 ± 0.180	0.459 ± 0.018	0.743 ± 0.019	0.959 ± 0.028	–	9.9 ± 0.2
17	7:36:47.552	+65:35:52.142	0.086	2.860 ± 0.326	1.943 ± 0.394	0.583 ± 0.077	0.892 ± 0.103	0.805 ± 0.086	0.795 ± 0.084	17.9 ± 1.7
18	7:36:47.200	+65:35:41.740	0.115	2.860 ± 0.193	3.699 ± 0.378	0.428 ± 0.034	0.818 ± 0.056	0.778 ± 0.050	1.137 ± 0.050	37.8 ± 2.1
19	7:36:47.114	+65:35:34.012	0.134	2.860 ± 0.413	2.272 ± 0.510	1.109 ± 0.182	0.815 ± 0.118	1.064 ± 0.144	0.169 ± 0.107	7.6 ± 0.9
20	7:36:46.617	+65:35:12.607	0.193	2.860 ± 0.319	2.091 ± 0.395	0.491 ± 0.063	0.527 ± 0.061	0.755 ± 0.078	0.321 ± 0.083	11.2 ± 1.0
21	7:36:45.999	+65:39:55.360	0.554	2.860 ± 0.722	2.232 ± 0.910	3.025 ± 0.862	0.239 ± 0.063	0.560 ± 0.131	0.585 ± 0.187	10.8 ± 2.2
22	7:36:41.780	+65:38:06.645	0.255	2.860 ± 0.087	1.787 ± 0.093	2.084 ± 0.072	0.264 ± 0.009	0.268 ± 0.011	0.066 ± 0.023	57.2 ± 1.4
23	7:36:39.878	+65:37:06.090	0.138	2.860 ± 0.168	1.954 ± 0.216	0.890 ± 0.060	0.567 ± 0.035	0.562 ± 0.032	0.119 ± 0.043	15.5 ± 0.7
24	7:36:43.136	+65:37:13.704	0.137	2.860 ± 0.208	3.837 ± 0.538	0.449 ± 0.038	0.656 ± 0.049	0.959 ± 0.065	0.434 ± 0.054	13.2 ± 0.8
25	7:36:44.902	+65:36:59.318	0.107	2.860 ± 0.078	3.091 ± 0.133	0.740 ± 0.023	0.513 ± 0.014	0.433 ± 0.013	0.197 ± 0.020	68.2 ± 1.5
26	7:36:46.581	+65:36:39.334	0.066	2.860 ± 0.137	3.918 ± 0.302	0.475 ± 0.026	0.723 ± 0.035	0.881 ± 0.039	0.628 ± 0.035	32.1 ± 1.3
27	7:36:47.883	+65:36:27.440	0.042	2.860 ± 0.319	4.171 ± 0.717	0.672 ± 0.088	0.761 ± 0.085	0.906 ± 0.096	0.761 ± 0.083	14.1 ± 1.3
28	7:36:53.405	+65:35:28.344	0.095	2.860 ± 0.256	3.081 ± 0.524	0.367 ± 0.041	0.826 ± 0.075	0.853 ± 0.072	0.946 ± 0.066	21.7 ± 1.6
29	7:36:55.716	+65:35:06.528	0.140	2.860 ± 0.364	4.103 ± 0.844	0.427 ± 0.067	0.753 ± 0.098	1.189 ± 0.146	0.115 ± 0.094	5.4 ± 0.6
30	7:36:56.334	+65:34:59.811	0.154	2.860 ± 0.472	–	0.482 ± 0.091	0.927 ± 0.154	0.993 ± 0.153	0.809 ± 0.122	9.9 ± 1.3
31	7:36:53.279	+65:36:31.060	0.075	2.860 ± 0.347	4.779 ± 0.908	0.394 ± 0.057	0.701 ± 0.086	1.044 ± 0.117	1.129 ± 0.090	27.6 ± 2.8
32	7:36:52.236	+65:36:21.969	0.041	2.860 ± 0.195	2.085 ± 0.225	0.310 ± 0.026	0.793 ± 0.055	0.757 ± 0.049	0.173 ± 0.051	24.5 ± 1.4
33	7:36:48.761	+65:35:55.652	0.063	2.860 ± 0.325	2.823 ± 0.517	0.412 ± 0.056	1.022 ± 0.117	1.121 ± 0.117	0.582 ± 0.084	18.8 ± 1.8
34	7:36:46.985	+65:35:41.295	0.118	2.738 ± 0.049	2.062 ± 0.169	0.281 ± 0.011	0.935 ± 0.021	0.831 ± 0.022	–	17.9 ± 0.3
35	7:36:57.586	+65:37:22.390	0.250	2.785 ± 0.075	4.692 ± 0.203	0.610 ± 0.019	0.597 ± 0.018	0.824 ± 0.027	–	9.7 ± 0.3
36	7:37:05.433	+65:36:42.233	0.250	2.860 ± 0.070	2.593 ± 0.124	1.580 ± 0.044	0.489 ± 0.012	0.692 ± 0.017	0.246 ± 0.018	51.9 ± 1.0
37	7:37:07.011	+65:36:36.410	0.259	2.704 ± 0.008	2.689 ± 0.035	1.713 ± 0.008	0.406 ± 0.002	0.461 ± 0.009	–	205.0 ± 0.6
38	7:37:18.452	+65:35:42.768	0.330	2.860 ± 0.149	2.971 ± 0.289	1.920 ± 0.113	0.514 ± 0.027	0.593 ± 0.030	0.018 ± 0.039	13.5 ± 0.6

Table 2. *Cont.*

ID	R.A. (J2000)	Dec. (J2000)	R/R_{25}	Hα	[O II]λ3727	[O III]λ5007	[N II]λ6583	[S II]λλ6717,31	A_V (mag)	Hβ 10^{-15}(erg s^{-1} cm^{-2})
(1)	(2)	(3)	(4)	(5)	(6)	(7)	(8)	(9)	(10)	(11)
39	7:37:19.431	+65:35:37.799	0.338	2.860 ± 0.154	3.354 ± 0.283	1.625 ± 0.099	0.515 ± 0.029	0.647 ± 0.034	0.713 ± 0.040	26.7 ± 1.2
40	7:36:50.723	+65:35:00.452	0.186	2.824 ± 0.024	1.532 ± 0.101	1.014 ± 0.011	0.461 ± 0.008	0.443 ± 0.013	–	17.1 ± 0.1
41	7:36:35.603	+65:36:48.393	0.174	2.860 ± 0.399	–	1.044 ± 0.166	0.559 ± 0.079	0.659 ± 0.088	0.231 ± 0.103	13.8 ± 1.6
42	7:36:38.275	+65:36:33.111	0.149	2.777 ± 0.080	2.153 ± 0.207	1.151 ± 0.036	0.538 ± 0.027	0.768 ± 0.034	–	14.8 ± 0.4
43	7:36:41.882	+65:36:09.148	0.128	2.840 ± 0.120	1.611 ± 0.282	0.832 ± 0.040	0.646 ± 0.032	0.775 ± 0.038	–	11.6 ± 0.5
44	7:36:47.955	+65:35:35.674	0.121	2.860 ± 0.377	1.600 ± 0.360	0.523 ± 0.080	0.714 ± 0.095	0.516 ± 0.065	0.247 ± 0.098	22.0 ± 2.4
45	7:36:22.076	+65:38:56.202	0.382	2.860 ± 0.218	3.773 ± 0.474	1.184 ± 0.102	0.427 ± 0.034	0.679 ± 0.050	0.227 ± 0.056	23.1 ± 1.4
46	7:36:20.847	+65:38:03.288	0.339	2.860 ± 0.415	3.784 ± 0.867	1.852 ± 0.304	0.316 ± 0.047	0.484 ± 0.066	0.470 ± 0.108	24.4 ± 2.9
47	7:36:20.019	+65:37:31.521	0.345	2.827 ± 0.077	3.781 ± 0.242	0.754 ± 0.025	0.495 ± 0.021	0.842 ± 0.033	–	13.0 ± 0.3
48	7:36:19.348	+65:37:15.125	0.359	2.860 ± 0.113	3.270 ± 0.189	0.957 ± 0.043	0.448 ± 0.018	0.613 ± 0.028	0.134 ± 0.029	66.3 ± 2.2
49	7:36:19.328	+65:37:05.527	0.366	2.826 ± 0.009	2.308 ± 0.030	1.850 ± 0.009	0.307 ± 0.001	0.292 ± 0.009	–	199.0 ± 0.6
50	7:37:18.249	+65:33:49.977	0.334	2.860 ± 0.077	2.395 ± 0.104	2.013 ± 0.061	0.248 ± 0.007	0.355 ± 0.012	0.535 ± 0.020	124.0 ± 2.8
51	7:37:17.668	+65:33:35.847	0.350	2.860 ± 0.220	2.288 ± 0.307	1.006 ± 0.088	0.342 ± 0.030	0.655 ± 0.052	0.043 ± 0.057	12.2 ± 0.8
52	7:37:17.171	+65:33:17.894	0.377	2.521 ± 0.101	2.599 ± 0.231	4.768 ± 0.189	0.294 ± 0.020	0.531 ± 0.037	–	7.8 ± 0.3
53	7:37:15.514	+65:32:05.246	0.524	2.777 ± 0.038	3.783 ± 0.140	1.324 ± 0.019	0.323 ± 0.008	0.788 ± 0.020	–	20.0 ± 0.3
54	7:37:15.053	+65:31:52.534	0.554	2.762 ± 0.071	3.885 ± 0.243	1.680 ± 0.044	0.282 ± 0.015	0.521 ± 0.026	–	11.7 ± 0.3
55	7:36:35.434	+65:35:08.504	0.334	2.860 ± 0.110	3.844 ± 0.230	1.285 ± 0.056	0.449 ± 0.018	0.633 ± 0.025	0.386 ± 0.029	36.7 ± 1.2
56	7:36:44.324	+65:35:04.887	0.237	2.860 ± 0.181	3.610 ± 0.335	1.086 ± 0.078	0.586 ± 0.038	0.793 ± 0.049	0.322 ± 0.047	18.1 ± 0.9
57	7:36:48.799	+65:35:00.381	0.203	2.860 ± 0.443	4.695 ± 1.121	1.087 ± 0.192	0.641 ± 0.101	1.053 ± 0.152	0.908 ± 0.115	9.0 ± 1.2
58	7:36:50.864	+65:34:59.651	0.187	2.860 ± 0.243	1.904 ± 0.288	1.114 ± 0.108	0.557 ± 0.049	0.650 ± 0.052	0.460 ± 0.063	11.9 ± 0.8
59	7:36:55.474	+65:34:09.316	0.287	2.860 ± 0.315	3.012 ± 0.498	0.967 ± 0.121	0.535 ± 0.061	0.907 ± 0.097	0.079 ± 0.082	8.8 ± 0.8
60	7:37:11.345	+65:33:56.794	0.292	2.860 ± 0.280	4.006 ± 0.578	1.374 ± 0.153	0.456 ± 0.046	0.816 ± 0.076	0.625 ± 0.073	17.1 ± 1.4
61	7:37:16.617	+65:33:51.188	0.323	2.860 ± 0.385	4.502 ± 1.027	0.783 ± 0.120	0.410 ± 0.057	0.877 ± 0.109	0.895 ± 0.100	13.2 ± 1.5
62	7:37:18.101	+65:33:51.336	0.332	2.860 ± 0.088	2.665 ± 0.118	2.013 ± 0.070	0.253 ± 0.008	0.315 ± 0.013	0.371 ± 0.023	51.9 ± 1.3
63	7:37:19.463	+65:33:50.704	0.341	2.860 ± 0.177	2.634 ± 0.254	1.357 ± 0.096	0.369 ± 0.024	0.537 ± 0.032	0.314 ± 0.046	23.8 ± 1.2
64	7:36:28.529	+65:33:50.382	0.598	2.860 ± 0.066	2.633 ± 0.092	3.448 ± 0.091	0.173 ± 0.004	0.310 ± 0.010	0.090 ± 0.017	89.2 ± 1.7
65	7:37:10.039	+65:32:55.986	0.416	2.860 ± 0.245	3.895 ± 0.510	1.903 ± 0.184	0.410 ± 0.036	0.631 ± 0.052	0.595 ± 0.064	24.8 ± 1.8
66	7:36:48.187	+65:39:50.257	0.555	2.860 ± 0.198	2.440 ± 0.259	3.278 ± 0.255	0.147 ± 0.013	0.335 ± 0.024	0.188 ± 0.051	24.9 ± 1.4
67	7:36:41.163	+65:39:32.192	0.466	2.860 ± 0.291	3.236 ± 0.463	2.884 ± 0.331	0.223 ± 0.024	0.382 ± 0.037	0.104 ± 0.075	17.8 ± 1.5
68	7:36:39.314	+65:39:27.757	0.446	2.860 ± 0.208	3.157 ± 0.360	2.713 ± 0.222	0.237 ± 0.019	0.428 ± 0.031	0.051 ± 0.054	19.6 ± 1.2
69	7:36:26.635	+65:38:53.255	0.359	2.860 ± 0.491	–	1.060 ± 0.206	0.514 ± 0.092	0.895 ± 0.164	0.182 ± 0.127	4.9 ± 0.7
70	7:36:25.091	+65:38:50.765	0.361	2.860 ± 0.184	2.989 ± 0.374	1.429 ± 0.104	0.414 ± 0.028	0.696 ± 0.044	0.490 ± 0.048	23.0 ± 1.2
71	7:36:23.743	+65:38:45.612	0.359	2.860 ± 0.289	2.961 ± 0.465	1.322 ± 0.151	0.537 ± 0.059	0.777 ± 0.075	0.034 ± 0.075	7.3 ± 0.6
72	7:36:48.963	+65:39:18.487	0.478	2.860 ± 0.182	2.953 ± 0.279	2.281 ± 0.164	0.255 ± 0.017	0.360 ± 0.025	0.456 ± 0.047	49.5 ± 2.6
73	7:36:36.239	+65:38:57.847	0.363	2.688 ± 0.085	3.557 ± 0.286	1.022 ± 0.034	0.623 ± 0.026	0.809 ± 0.042	–	9.3 ± 0.3
74	7:36:34.204	+65:38:53.771	0.351	2.706 ± 0.080	–	1.151 ± 0.038	0.478 ± 0.019	1.020 ± 0.040	–	8.4 ± 0.2
75	7:36:17.843	+65:37:30.808	0.370	2.860 ± 0.198	3.711 ± 0.435	0.963 ± 0.077	0.410 ± 0.029	0.709 ± 0.048	0.167 ± 0.051	42.7 ± 2.4
76	7:36:19.527	+65:37:08.371	0.361	2.687 ± 0.015	2.442 ± 0.040	1.722 ± 0.010	0.332 ± 0.003	0.311 ± 0.009	–	178.0 ± 0.9
77	7:36:23.465	+65:36:19.340	0.361	2.860 ± 0.773	4.711 ± 2.049	1.598 ± 0.488	0.445 ± 0.123	0.704 ± 0.178	0.820 ± 0.200	17.1 ± 3.8

Table 2. *Cont.*

ID	R.A. (J2000)	Dec. (J2000)	R/R_{25}	Hα	[O II]λ3727	[O III]λ5007	[N II]λ6583	[S II]λλ6717,31	A_V (mag)	Hβ 10^{-15} (erg s^{-1} cm^{-2})
(1)	(2)	(3)	(4)	(5)	(6)	(7)	(8)	(9)	(10)	(11)
78	7:36:24.820	+65:36:02.233	0.369	2.837 ± 0.205	3.446 ± 0.705	1.559 ± 0.109	0.432 ± 0.054	0.980 ± 0.097	–	5.8 ± 0.4
79	7:36:27.838	+65:35:28.437	0.389	2.788 ± 0.038	3.001 ± 0.284	0.392 ± 0.010	0.475 ± 0.009	0.528 ± 0.019	–	27.4 ± 0.3
80	7:36:39.408	+65:37:10.680	0.147	2.860 ± 0.348	3.802 ± 0.738	1.278 ± 0.176	0.511 ± 0.063	0.482 ± 0.055	0.613 ± 0.090	16.5 ± 1.7
81	7:36:45.748	+65:36:57.650	0.102	2.837 ± 0.029	1.868 ± 0.036	0.988 ± 0.011	0.438 ± 0.005	0.328 ± 0.006	–	110.0 ± 1.1
82	7:36:49.018	+65:36:52.259	0.097	2.860 ± 0.088	2.670 ± 0.179	1.092 ± 0.039	0.531 ± 0.017	0.487 ± 0.017	0.948 ± 0.023	75.5 ± 1.9
83	7:36:51.827	+65:36:45.214	0.097	2.822 ± 0.015	1.891 ± 0.026	1.295 ± 0.008	0.428 ± 0.003	0.366 ± 0.008	–	102.0 ± 0.5
84	7:36:59.802	+65:36:33.455	0.159	2.860 ± 0.554	2.063 ± 0.633	0.755 ± 0.167	0.576 ± 0.113	0.989 ± 0.178	0.708 ± 0.144	12.1 ± 1.9

Note. Columns: (1) Object number of the H II region; (2)–(3) object coordinate in J2000; (4) galactocentric distance normalized by R_{25}; (5)–(9) emission-line fluxes normalized by Hβ; (10) dust attenuation at the V band; (11) the fluxes of Hβin units of 10^{-15} erg s^{-1} cm^{-2}.

2.4. Ancillary Data from the Literature

Complementary with the observed spectra, we additionally compiled a dataset of spectral emission lines for 104 H II regions in NGC 2403 from the literature, all of which have been corrected for dust attenuation. Specifically, the supplementary sample contained 6 H II regions collected from McCall et al. [33], 5 H II regions collected from Fierro et al. [34], 12 H II regions collected from Garnett et al. [35], 9 H II regions collected from Bresolin et al. [36], 3 H II regions collected from Garnett et al. [37], 17 H II regions collected from van Zee et al. [38], 7 H II regions collected from Berg et al. [39], 12 H II regions collected from Mao et al. [18], and 33 H II regions collected from Rogers et al. [40]. The spatial locations for these H II regions along with the observed ones are all shown in Figure 1.

For the common H II regions observed in this work and collected from other studies, we compared the emission-line fluxes, and the result is shown in Figure 2. Most of the data points were enveloped within the $\pm 2\sigma$ (± 0.24 dex) demarcation, which verified the consistency of the data. Differences between the apertures in size and position for extracting the spectra for the same H II regions in the different studies were considered to be a general source of the deviation. Specifically, a few of [O II]λ3727 and [S II]$\lambda\lambda$6717, 6731 lines deviated to a relatively large extent, which was ascribed to influences of different noise levels at the blue end of the spectra on [O II]λ3727 and different deblending qualities for [S II]$\lambda\lambda$6717, 6731 doublets, respectively.

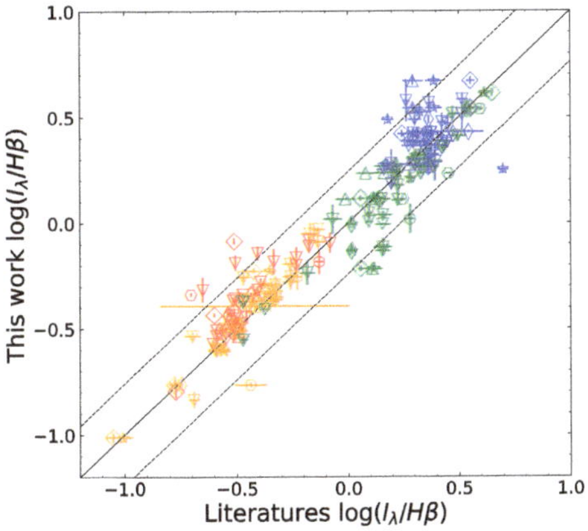

Figure 2. Comparison between the emission-line fluxes for the same H II regions measured in this work and those quoted from other studies, including McCall et al. [33], Fierro et al. [34], Garnett et al. [35], Bresolin et al. [36], Garnett et al. [37], Mao et al. [18], and Rogers et al. [40]. The emission lines include [O II] λ3727 (blue), [O III] $\lambda\lambda$4959,5007 (green), [N II] λ6584 (yellow), and [S II] $\lambda\lambda$6717,6731 (red), and they are normalized by Hβ. The data points are marked using the same symbols as in Figure 1. The line of unity is represented as the black solid line. The $\pm 2\sigma$ (± 0.24 dex, i.e., the standard deviation of the data) range are enveloped by the black dotted lines.

3. Photoionization Model

We adopted model grids developed by Levesque et al. [12], which combine the STARBURST99 stellar population synthesis model (Leitherer et al. [41], Vázquez and Leitherer [42]) with the MAPPINGS III photoionization model (Sutherland and Dopita [43], Allen et al. [44]). In this work, ionizing stellar populations were modeled by assuming star formation history of an instantaneous burst in the range 0–9.5 Myr with the step length 0.5 Myr in age, 5 metallicity values including $Z = 0.001, 0.004, 0.008, 0.02$, and 0.04 (corresponding to 0.05, 0.2, 0.4, 1, and 2 $Z_{\odot}$), and the Geneva "High" mass-loss star evolutionary tracks (Schaller et al. [45], Schaerer et al. [46], Charbonnel et al. [47], Schaerer et al. [48]; recommended in Levesque et al. [12]).

Ionized gaseous nebulae were configured in a parallel-plane geometry. The ionization parameter q at the inner surface of the nebulae was set to spanning $q = 1 \times 10^7, 2 \times 10^7, 4 \times 10^7, 8 \times 10^7, 1 \times 10^8, 2 \times 10^8, 4 \times 10^8$ cm s^{-1}. The gas-phase metallicity matched with that of the model stellar populations. The electron density $n_e = 100\,\text{cm}^{-3}$ was chosen in accordance with the low-density limit for H II regions in NGC 2403 (Garnett et al. [35], Berg et al. [39], Rogers et al. [40]).

4. Results

Strong emission lines in optical spectra for ionized gaseous nebulae encode a wealth of information and serve as effective diagnostics for a series of physical parameters. In this section, we present results of diagnoses from a series of strong-line indices for the H II regions in NGC 2403 and in turn inspect influences of underlying parameters on the emission lines.

4.1. BPT Diagram

The diagnostic diagram of the indices [O III]λ5007/Hβ versus (vs.) [N II]λ6583/Hβ proposed by Baldwin, Phillips, and Terlevich (denoted as BPT, Baldwin et al. [49]) is a powerful tool for classifying emission-line objects into active galactic nuclei (AGNs), low-ionization-emission regions (LINERs), or star-forming regions (Kewley et al. [50], Kauffmann et al. [51]). Figure 3 is the BPT diagram for the H II regions in NGC 2403, with the 0.0 Myr model grid as well as the Kewley et al. [50] and Kauffmann et al. [51] demarcation curves for the classification superimposed. Almost all of the points are consistent with the Kauffmann et al. [51] curve, and only one falls between the Kauffmann et al. [51] and Kewley et al. [50] curves, indicating that nearly all samples originate from star formation activity. By comparing the color-coded data points with the model grid, the data locus in the BPT diagram reflects a sequence of metallicity or/and the ionization parameter varying with the galactocentric distance and suggests a correlation of the two indices with both metallicity and the ionization parameter.

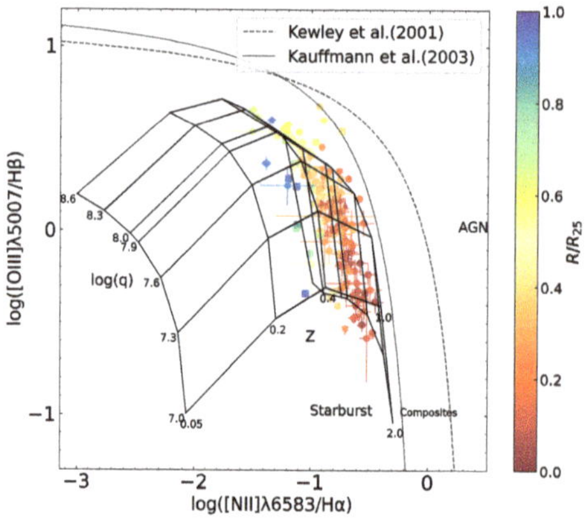

Figure 3. log([O III]λ5007/Hβ) as a function of log([N II]λ6583/Hβ) (the BPT diagram) for the H II regions in NGC 2403 with the Levesque et al. [12] model grid (the constant age 0 Myr, variations in the metallicity Z (in units of $Z_\odot$), and the ionization parameter q) superimposed. The data points are marked using the same symbols as in Figure 1 and color-coded by the galactocentric distance (de-projected) in units of R_{25}. The criteria for classifying ionizing sources (stars and AGNs) are quoted from Kewley et al. [50] (black dotted line) and Kauffmann et al. [51] (black solid line).

4.2. Metallicity Estimates

The oxygen abundance for the H II regions in NGC 2403 was estimated for assessing the metallicity. There are a number of recipes for obtaining the oxygen abundance but considerable discrepancies among them, leaving an important challenge in metallicity studies (Maiolino and Mannucci [2], Kewley et al. [3], Mao et al. [18], Kewley and Ellison [52]). In this work, we adopted four widely used empirical diagnostics for estimating oxygen abundance. They are described as follows.

*R*23 (Pilyugin and Thuan [7]):[3]

$$12 + \log(O/H)_{high} = \frac{R23 + 726.1 + 842.2P + 337.5P^2}{85.96 + 82.76P + 43.98P^2 + 1.793R23}, \tag{3}$$

where $R23 \equiv \log \frac{[\text{O II}]\lambda3727 + [\text{O III}]\lambda\lambda4959,5007}{\text{H}\beta}$, $P = \frac{[\text{O III}]\lambda\lambda4959,5007}{[\text{O III}]\lambda\lambda4959,5007 + [\text{O II}]\lambda3727}$;

*O*3*N*2 (Pettini and Pagel [6]):

$$12 + \log(O/H) = 8.73 - 0.32O3N2, \tag{4}$$

where $O3N2 \equiv \log \frac{[\text{O III}]\lambda5007/\text{H}\beta}{[\text{N II}]\lambda6583/\text{H}\alpha}$;

*N*2 (Pettini and Pagel [6]):

$$12 + \log(O/H) = 9.37 + 2.03 \times N2 + 1.26 \times N2^2 + 0.32 \times N2^3, \tag{5}$$

where $N2 \equiv \log \frac{[\text{N II}]\lambda6583}{\text{H}\alpha}$;

*N*2*O*2 (Bresolin [8]):

$$12 + \log(O/H) = 8.66 + 0.36 \times N2O2 - 0.17 \times N2O2^2, \tag{6}$$

where $N2O2 \equiv \log \frac{[\text{N II}]\lambda6583}{[\text{O II}]\lambda3727}$.

From the above diagnostics, we estimated the oxygen abundances for the H II regions in NGC 2403, whose radial profiles are displayed in Figure 4, with the best-fit linear curves superimposed. The uncertainty of the slope was quantified as a standard deviation derived from bootstrapping the sample 3000 times, and the corresponding fit parameters are listed in Table 3. A negative gradient is obviously seen from each of the profiles, consistent with the inside-out scenario of galaxy evolution. However, differences in gradient slope and data dispersion between these radial profiles are evident and imply greater complexity for the emission-line indices, which inevitably affects and biases relevant studies based on metallicity gradients derived from these recipes.

Table 3. Fitting parameters of radial metallicity distribution.

Indicator (1)	Gradient (dex/R_{25}) (2)	σ (dex) (3)	R^2 (4)
$R23_{P05}$	-0.31 ± 0.04	0.17	0.35
$O3N2_{P04}$	-0.35 ± 0.05	0.12	0.41
$N2_{P04}$	-0.31 ± 0.03	0.09	0.58
$N2O2_{B07}$	-0.58 ± 0.02	0.08	0.81

Notes. Columns: (1) Metallicity indicator; (2)–(4) slope of the radial metallicity gradient, dispersion, and coefficient of determination of the fit, as presented in Figure 4.

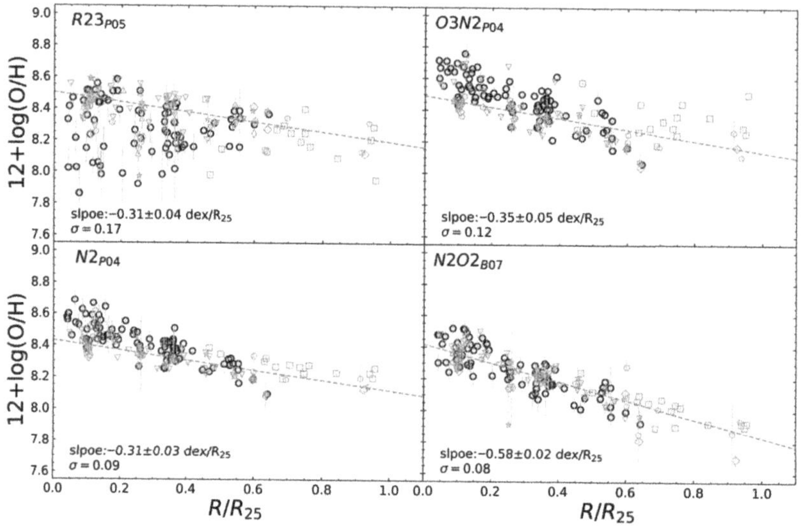

Figure 4. Radial distributions of the oxygen abundance derived from the indices *R*23 ((**top-left panel**), the calibration is addressed in Pilyugin and Thuan [7] and denoted as $R23_{\rm P05}$), *O*3*N*2 ((**top-right panel**), the calibration is addressed in Pettini and Pagel [6] and denoted as $O3N2_{\rm P04}$), *N*2 ((**bottom-left panel**), the calibration is addressed in Pettini and Pagel [6] and denoted as $N2_{\rm P04}$), and *N*2*O*2 ((**bottom-right panel**), the calibration is addressed in Bresolin [8] and denoted as $N2O2_{\rm B07}$) for the H II regions (represented by the same symbols as assigned in Figure 1) in NGC 2403 with the weighted linear best-fit curves (cyan dashed lines) superimposed. The fitting parameters are compiled in Table 3.

4.3. Index–Index Diagrams

In order to inspect the emission-line indices in more depth, we carried out a series of diagnoses by diagramming relations of the indices for the data in combination with the model grids. These diagnoses result in a further understanding of not only underlying parameters in the observed features but also more properties for the H II regions in NGC 2403. In this section, the emission-line indices inspected in the diagnoses included the metallicity indicators *R*23, *O*3*N*2, *N*2, and *N*2*O*2. As a widely used proxy of the ionization parameter, *O*3*O*2 ($\equiv$ [O III]$\lambda\lambda$4959, 5007/[O II]λ3727) was also employed in our work for specially tracing ionization states.[4]

4.3.1. Diagnosis at Fixed Age

Figure 5 shows the interrelations between *R*23, *O*3*N*2, *N*2, *N*2*O*2, and *O*3*O*2, with the model grid reflecting the initial situation (i.e., the 0 Myr age) superimposed in each of the panels. As illustrated in these diagrams, among the four metallicity indicators, the correlation with *O*3*O*2 is strong for *O*3*N*2 and *N*2 (Pearson coefficients $\rho = 0.88$ and -0.72, respectively), moderate for *R*23 ($\rho = 0.45$), and weak for *N*2*O*2 ($\rho = -0.39$). Different levels of degeneracy of metallicity and the ionization parameter in the different indices are disclosed by the model grids. Specifically, for constant values of metallicity, *O*3*N*2 increases monotonically with the ionization parameter, whereas *N*2 decreases at higher ionization states with a slight curvature; dependence of *R*23 on the ionization parameter appears complicated, with two opposite trends to the low- and high-metallicity ends, similar to the so-called double-value effect of *R*23 on its estimates of the oxygen abundance; by contrast, *N*2*O*2 stays robust against variations in the ionization parameter, as manifested by the extensive model grid and the widespread data distribution in the *O*3*O*2-*N*2*O*2 diagram in

Figure 5. This result verifies the non-negligible sensitivities of *O3N2* and *N2* to the ionization parameter, as well as the reliability of *N2O2* for estimating metallicity, particularly at variable ionization states. The sensitivity of *R23* to the ionization parameter was inherently weaker than that of *O3N2* and *N2*; furthermore, the correction for the ionization influence was taken into account in the *R23* calibration (Pilyugin and Thuan [7]).

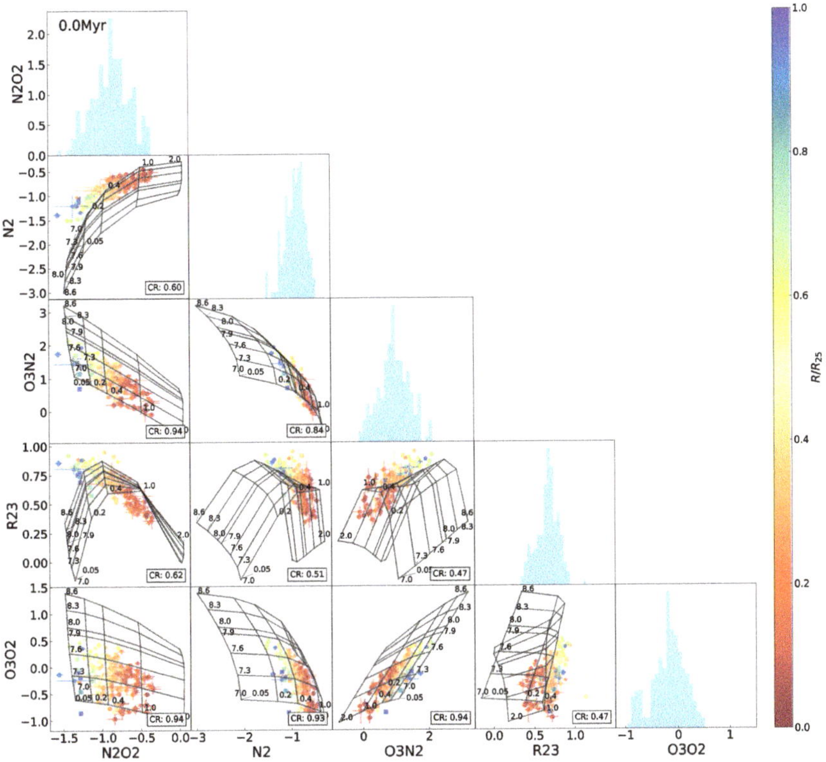

Figure 5. Interrelations between every two indices among *R23*, *O3N2*, *N2*, *N2O2*, *O3O2* for the H II regions in NGC 2403, with the Levesque et al. [12] model grid (the constant age of 0 Myr, variations in metallicity Z (in units of $Z_{\odot}$) and the ionization parameter q) superimposed. The data points are marked using the same symbols as in Figure 1 and color-coded the same as in Figure 3. The rate of the data covered by the model grid (defining the number ratio of the data points falling within the grid to the total of the data points, denoted as the coverage rate or "CR") is presented in the lower-right corner in each of the panels. Histograms of the indices are also shown.

Nonetheless, the data points from the outer part of the galaxy ($r > 0.2R_{25}$) lie in the conjunction zone of the higher and lower branches of the *R23* calibration and thus appear with dispersion in the *R23*-based estimates of metallicity. For the H II regions located in the inner area of the galaxy ($r \lesssim 0.2R_{25}$), even larger dispersion emerges in the *R23*-based estimates of metallicity (see the top-left panel of Figure 4). This is likely due to an inherent problem in the Pilyugin and Thuan [7] calibration. We found these data points in possession of lower excitation levels (quantified by the excitation parameter *P* as defined in Equation (3)). As pointed out in Moustakas et al. [56], the R23 calibration lacks sufficient low-excitation ($P \lesssim 0.2$) data, which prevents its good performance in such a regime and thereby induces the dispersion in estimates of metallicity.[5]

The *N2O2* metallicity diagnostic is based on the relation of N/O–O/H at high metallicity, where secondary nitrogen production leads to a correlation between N/O and O/H

(Maiolino and Mannucci [2], Kewley and Dopita [9]). For NGC 2403, Berg et al. [39] found that within $R \sim 0.93\,R_{25}$ (our estimate), $\log(N/O)$ and $12 + \log(O/H)$ exhibited a tight linear correlation, without the plateau characteristic of primary nitrogen production. Together with its insensitivity to the ionization parameter and immunity to the double-valued degeneracy, this supports the robustness of *N2O2* as a metallicity indicator.

As a consequence, among the four panels in Figure 4, the metallicity estimated from *N2O2* shows the most reliable radial metallicity gradient for NGC 2403, with a slope of -0.58 ± 0.02 dex R_{25}^{-1}, a dispersion of $\sigma = 0.08$ dex, and a coefficient of determination of $R^2 = 0.81$.

Comparison of the data distribution with the model grid in each of the index–index diagrams in Figure 5 manifests different extents of coverage of the model to the data points. Specifically, there are approximately 93–94% of the data points covered by the model grid in each of the *O3N2-N2O2*, *O3O2-N2*, and *O3O2-O3N2* diagrams, about 84% in the *O3N2–N2* diagram, and around 60% or even less in the *N2O2-N2* and *R23*-relative diagrams. The model appears to characterize the data well in the *O3N2-N2O2*, *O3O2-N2*, *O3O2-O3N2*, and *O3N2–N2* diagrams but not in the others. In each of the *N2O2-N2* and *R23*-relative diagrams, the data locus fails to follow the bending regime of the model grid but keeps a more linear trend in the plane. As a result, some of the data points exceed the model coverage, and they are the H II regions located in the outskirts of NGC 2403.

4.3.2. Diagnosis with Age Evolution

The model grids in Figures 3 and 5 were established at the fixed age of 0 Myr (in an initial situation) for the purpose of eliminating effects of age but specially focusing on behaviors of metallicity and the ionization parameter. As a matter of fact, emission lines are supposed to tightly relate with the age of ionizing stellar populations, due to a decrease in the number of ionizing photons when stellar populations age. With the diagrammatic diagnosis independent of age worked out above, we conducted modeling of the spectral indices by varying the age (range of 0–9.5 Myr) in this section, to inspect the age evolution of the index–index relationships, whose results are shown in Figures 6–9 and Appendix A. The coverage rates for all diagnostic diagrams are summarized in Table 4.

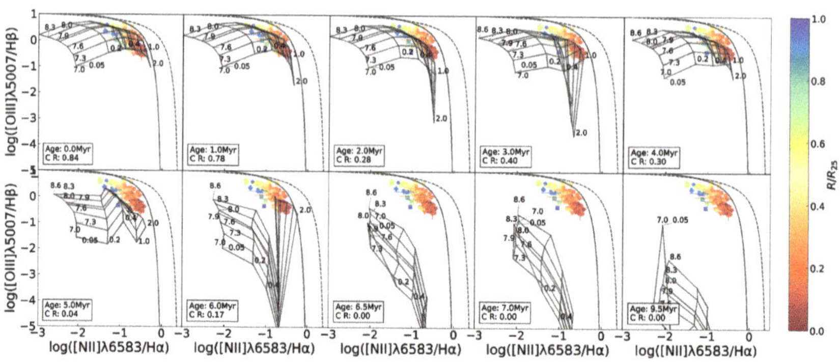

Figure 6. BPT diagrams for the H II regions in NGC 2403 with the Levesque et al. [12] model grids (constant age, variations in metallicity Z (in units of $Z_{\odot}$) and the ionization parameter q in each of the panels) superimposed. The data points and the dotted and solid lines have the same meaning in all of the panels as in Figure 3). In the different panels, the model grids vary with the age ranging from 0 to 9.5 Myr (This only shows part of the age grid as a representative.) The age and the CR (the same definition as in Figure 5) are listed in the lower-right corner in each of panels. The data points are marked using the same symbols as in Figure 1 and color-coded the same as in Figure 3.

Table 4. Coverage rates of diagnostic diagram.

AGE (Myr) (1)	BPT (2)	$N2O2-N2$ (3)	$N2O2-O3N2$ (4)	$N2O2-R23$ (5)	$N2O2-O3O2$ (6)	$N2-O3N2$ (7)	$N2-R23$ (8)	$N2-O3O2$ (9)	$O3N2-R23$ (10)	$O3N2-O3O2$ (11)	$R23-O3O2$ (12)	[N II]-[O III] (13)	[N II]-[O II] (14)	[O II]-[O III] (15)	Average (16)
0	0.84	0.60	0.94	0.62	0.94	0.84	0.51	0.93	0.47	0.94	0.47	0.84	0.60	0.46	0.71
0.5	0.84	0.60	0.95	0.60	0.95	0.84	0.49	0.95	0.46	0.95	0.47	0.83	0.60	0.45	0.71
1	0.78	0.60	0.98	0.61	0.97	0.78	0.44	0.95	0.44	0.97	0.47	0.78	0.59	0.45	0.70
1.5	0.51	0.59	0.98	0.52	0.98	0.51	0.37	0.85	0.39	0.98	0.45	0.54	0.59	0.44	0.62
2	0.28	0.63	0.97	0.23	0.98	0.28	0.29	0.77	0.44	0.98	0.51	0.28	0.64	0.49	0.55
2.5	0.07	0.55	0.86	0.05	0.86	0.07	0.11	0.40	0.34	0.85	0.43	0.08	0.57	0.43	0.40
3	0.40	0.57	0.99	0.09	0.99	0.40	0.16	0.93	0.22	0.99	0.29	0.42	0.58	0.29	0.52
3.5	0.19	0.54	0.99	0.10	0.99	0.19	0.12	0.91	0.26	0.99	0.36	0.21	0.55	0.33	0.48
4	0.30	0.55	0.99	0.07	0.99	0.30	0.16	0.94	0.24	0.99	0.31	0.32	0.55	0.28	0.50
4.5	0.33	0.44	0.81	0.06	0.80	0.33	0.09	0.85	0.15	0.80	0.15	0.35	0.48	0.15	0.41
5	0.04	0.27	0.80	0.00	0.78	0.04	0.00	0.51	0.15	0.81	0.17	0.04	0.30	0.15	0.29
5.5	0.11	0.20	0.25	0.00	0.24	0.11	0.00	0.40	0.15	0.24	0.10	0.13	0.23	0.15	0.17
6	0.17	0.26	0.01	0.00	0.01	0.17	0.00	0.31	0.12	0.01	0.05	0.17	0.27	0.03	0.11
6.5	0.00	0.27	0.00	0.00	0.00	0.00	0.00	0.00	0.04	0.00	0.00	0.00	0.27	0.00	0.04
7	0.00	0.24	0.00	0.00	0.00	0.00	0.00	0.00	0.02	0.00	0.00	0.00	0.26	0.00	0.04
7.5	0.00	0.25	0.00	0.00	0.00	0.00	0.00	0.00	0.00	0.00	0.00	0.00	0.27	0.00	0.04
8	0.00	0.22	0.00	0.00	0.00	0.00	0.00	0.00	0.00	0.00	0.00	0.00	0.22	0.00	0.03
8.5	0.00	0.18	0.01	0.00	0.01	0.00	0.00	0.00	0.19	0.01	0.19	0.00	0.17	0.00	0.05
9	0.00	0.16	0.00	0.00	0.00	0.00	0.00	0.00	0.00	0.00	0.00	0.00	0.16	0.00	0.02
9.5	0.00	0.05	0.00	0.00	0.00	0.00	0.00	0.00	0.00	0.00	0.00	0.00	0.05	0.00	0.01

Notes. Columns: (1) Age of stellar population; (2)–(16) CR of diagnostic diagrams.

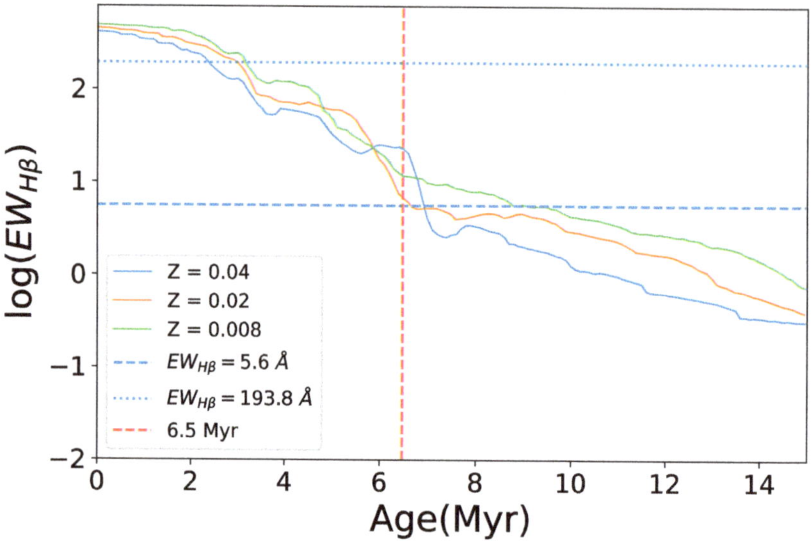

Figure 7. Constraints on the stellar population age derived from the equivalent width of the Hβ emission line ($EW_{\mathrm{H}\beta}$). The model curves show $EW_{\mathrm{H}\beta}$–age relations for instantaneous bursts at different metallicities from the STARBURST99 model (Leitherer et al. [41]) using the publicly available 1999 dataset downloaded from the official website. Horizontal dotted and dashed lines mark the observed maximum and minimum $EW_{\mathrm{H}\beta}$ values of the emission lines across the galaxy. The red vertical line indicates the 6.5 Myr upper-age limit inferred from the BPT diagram coverage. The minimum $EW_{\mathrm{H}\beta}$ value is found in the central, higher-metallicity region, consistent with an age of ~7 Myr (Z = 0.02).

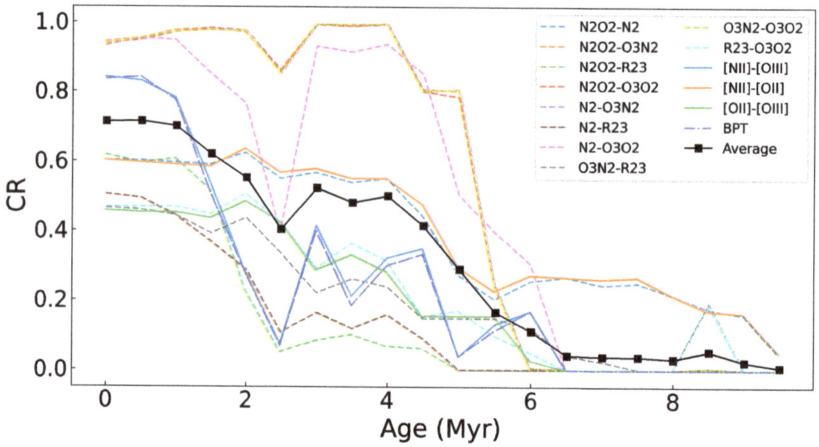

Figure 8. The CR (the same definition as in Figure 5) of the model grid as a function of the age ranging from 0 to 9.5 Myr, for the diagrams in Figures 3, 5 and 10. The dashed lines are color-coded by specific index–index relations with the legend displayed in the top-right corner. The black filled boxes connected by the black solid line represent the average evolution of the CR among all of the diagrams.

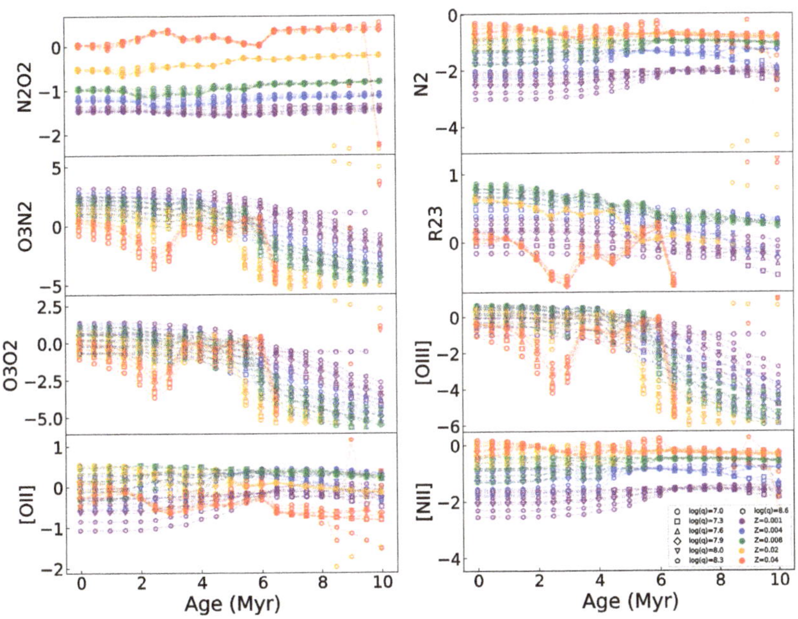

Figure 9. Spectral index or Hβ-normalized emission lines of [O III]λ5007, [N II]λ6583, and [O II]λ3727 (hereafter referred to as [O III], [N II], and [O II], respectively) as a function of age reproduced by the Levesque et al. [12] model. In each of the panels, the model products are color-coded by metallicity and symbol-coded by the ionization parameter, with the legend listed in the lower-left corner in the bottom-right panel.

Figure 6 shows a modeled evolution of the BPT relationship superimposed on the observed data. The most apparent behavior of the model grid is a progressive downward shift with aging, i.e., an intensive decline in [O III]λ5007/Hβ relative to a weak change in [N II]λ6583/Hα. It is because the [O III] emission lines are more preferentially excited at a higher degree of ionization, in contrast to the [N II]λ6583 emission lines associated with low-ionization environments (Kewley et al. [3], Osterbrock and Ferland [32]); with the aging of stellar populations and resultant softening of stellar radiations, H II regions have lower ionization states and more delicate [O III] excitation. A direct aftermath of the change in the spectral features is a decrease in the coverage rate of the model to the data in the diagram. The model grids with 0–1 Myr cover ∼80% of the data points, but the coverage rate rapidly reduces to <30% at ≥2 Myr and 0% (no coverage) at ≥6.5 Myr. The total separation of the model grid from the observed data suggests the upper age limit ∼6.5 Myr for the H II regions in NGC 2403 under this framework. We further cross-checked the age constraint using the equivalent width of Hβ ($EW_{\mathrm{H}\beta}$) as predicted by the Starburst99 models for instantaneous bursts (Leitherer et al. [41]), based on the 1999 dataset from the official website[6]. For each metallicity, we compared the model $EW_{\mathrm{H}\beta}$–age relations with the observed maximum and minimum $EW_{\mathrm{H}\beta}$ values in Figure 7. The smallest $EW_{\mathrm{H}\beta}$ was found in the central high-metallicity regions, corresponding to an age of ∼7 Myr (for $Z = 0.02$) in the models. Given that the observed continuum contains contributions from the underlying stellar populations of the galactic disk in addition to the H IIregions, the measured $EW_{\mathrm{H}\beta}$ is likely underestimated, thus biasing the age estimate toward older values. Taking this effect into account, the upper limit of ∼6.5 Myr inferred from the BPT coverage remains a reasonable estimate.

Combining Figures 3, 5, 6 and A1–A9, we find that a few of the data points from the central area of the galaxy are missed by the 0 Myr model but covered by the 2–6 Myr grids, which indicates an older stage for these H II regions than others in the sample. It is interesting that at the 3–6 Myr evolution phase, the model grids emerge with peculiar fluctuation at the high-metallicity part, which is ascribed to an influence of Wolf–Rayet (W-R) stars on the [O III] emission lines (Levesque et al. [12]).

Variations in the model coverage to the observed data in the index–index diagrams with age are shown in Figure 8, with an evolutionary trend of the average coverage among all of the diagrams superimposed. We can see that in general, for almost all of the diagrams, the model coverage stands at the highest levels at 0–1 Myr but drops sharply at 2–6 Myr and is lowest after 6 Myr. The highest coverage rates differ between the various diagrams by ∼50% at most. All of the *R*23-relative diagrams possess the coverage rate ≤60%. It is suspected that the model grids are likely to have some problem in reproducing the [O II] emission line or/and sampling the parameter spaces, which is discussed in Section 5.2. By contrast, the *N*2*O*2-*O*3*N*2 and *N*2*O*2-*O*3*O*2 curves maintain high coverage rates for a long period (≥4 Myr timescale), because the ionization parameter and metallicity are more effectively decoupled in the two diagrams, leading to larger areas of the model grids with higher possibilities for covering the data points. After 6 Myr, the coverage rate for most of the diagrams decreases to <10%, except for *N*2*O*2-*N*2 and [N II]-[O II], which are able to keep ∼30% coverage rate after 6 Myr. This is because the [O II] and [N II] emission lines are low-ionization emission lines not as sensitive to age as others, which is depicted in Figure 9[7].

By modeling *N*2*O*2, *N*2, *O*3*N*2, *R*23, *O*3*O*2, [O II], [O III], and [N II] at a variety of values for metallicity, the ionization parameter, and stellar population age addressed in Levesque et al. [12], Figure 9 reproduces the age evolution of the spectral indices and the emission line ratios, at each constant metallicity and ionization parameter, respectively, as an underlying factor interpreting the above features in the diagrams (Figures 5, 6, 10 and A1–A18). We can see from the evolutionary tracks that the indices *N*2*O*2 and *N*2 are the least sensitive to age; *N*2*O*2 even appears invariant with respect to the ionization parameter in the whole age range of 0–10 Myr, suggesting negligible degeneracy (if any) of metallicity, ionization parameter, and stellar population age in these indices, but *N*2 jointly depends on both metallicity and ionization parameter for H II regions during the first 6 Myr.

The *O*3*N*2 tracks gradually decline with age at lower metallicity than the solar value ($Z \leq 0.02$) but fluctuate conspicuously in the 3–6 Myr range at higher metallicities ($Z = 0.04$); *O*3*N*2 is also sensitive to both metallicity and ionization parameter, similar to *N*2 but spanning the whole 0–10 Myr range. *O*3*O*2 is similar to *O*3*N*2 in terms of age tracks but more dependent on the ionization parameter than metallicity, and this is the reason for this index being commonly adopted as an ionization proxy. The *R*23 index exhibits a weak declining trend with age and a decreasing sensitivity to the ionization parameter as metallicity increases. The fluctuation in the 3–6 Myr range is a noticeable feature in all of the tracks for *O*3*N*2, *O*3*O*2, and *R*23 at the metallicity $Z = 0.04$, coincident with the [O III] tracks at the same age and metallicity. As explained above, this feature is caused by the influence of W-R stars on the [O III] emission line, which results in the peculiar behaviors of the 3–6 Myr model grids in all of the [O III] relative diagrams and has potential impact on observed data loci in this special age interval. The relation with age is strong for [O III] but weak for [O II] and [N II] emission lines, which suggests strong age dependence of *O*3*N*2 and *O*3*O*2 but insensitivity to *N*2*O*2 and *N*2.

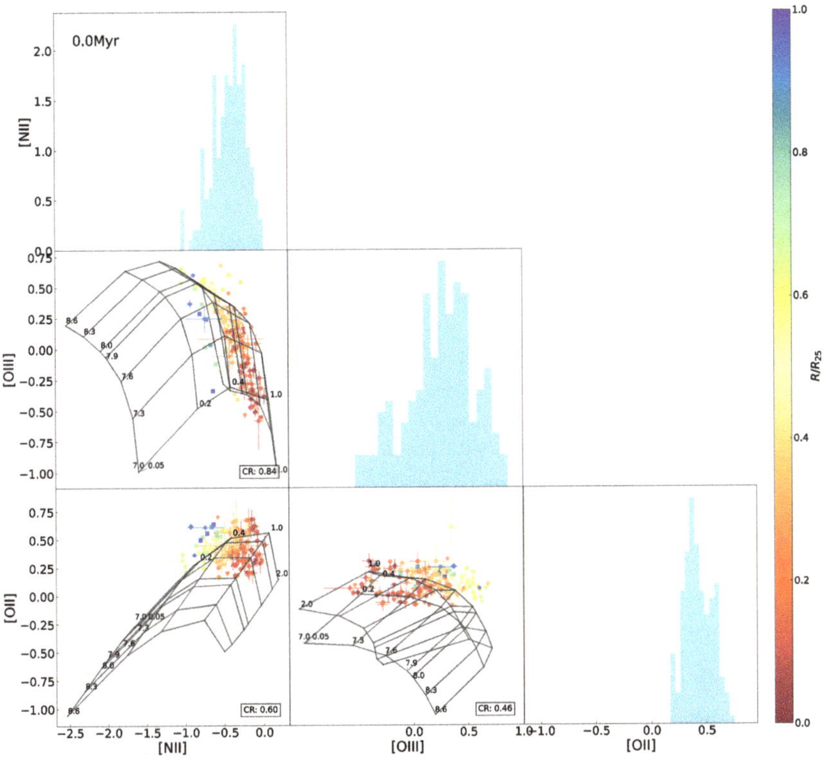

Figure 10. Interrelations between every two emission-line ratio (normalized by Hβ) among [O II], [N II], and [O III] for the H II regions in NGC 2403, with the Levesque et al. [12] model grid (constant age of 0 Myr, variations in metallicity Z (in units of $Z_{\odot}$) and the ionization parameter q) superimposed. The data points are marked using the same symbols as in Figure 1 and color-coded the same as in Figure 3. The CR (same definition as in Figure 5) is presented in the lower-right corner in each of the panels. Histograms of the indices are also shown.

5. Discussion

5.1. Effects of Apertures

Differences in position and size of apertures for spectral extraction have much potential to introduce extra uncertainties in the measurements. The specific effects of apertures on different emission lines and different spectral indices are suspected to be different. For instance, higher-ionization emission lines such as [O III] tend to be more compact and concentrated in central parts of H II regions, whereas lower-ionization emission lines such as [O II] and [N II] often extend over a wider scale (Kewley et al. [3]); for the behaviors of spectral indices, spatially resolved observations of large H II regions have revealed that *O3O2* peaks around the nebular centers, while *N2O2* is relatively invariable across different locations within the H II regions (Mao et al. [18]).

As a consequence, if apertures are more centrally placed, they are more likely to capture stronger [O III] emission lines, leading to higher values for *O3O2*. By contrast, larger apertures enclosing more outer areas of H II regions are more likely to extract stronger [O II] and [N II] emission lines and hence obtain a decrease in *O3O2* or *O3N2*.

5.2. Potential Limitation in the Model on the [O II] Emission Line

In the above results, the model fails to adequately fit the observed data in the *N2O2*-*N2* diagram as well as *R23*-relative diagrams, particularly for the H II regions located in outskirts of the galaxy with relatively low metallicities. In order to find an interpretation, we further inspected line–line diagrams, as shown in Figure 10. We can see that the model grid covers almost all of the data points in the [O III]-[N II] diagram but only one half in the [N II]–[O II] and [O III]–[O II] diagrams. Variations in age are sufficient to account for the low coverage in the [N II]–[O II] and [O III]–[O II] diagrams (see Figure 8).

The parameter space of metallicity and the ionization parameter in the model, in principle, is supposed to cover the observed data. It is unclear whether the failure of the coverage in the [O III]–[O II] diagram is caused by inadequate sampling resolution or other inherent limitations in the model. From the [N II]–[O II] diagram, we can affirm that the Levesque et al. [12] model grids are unable to reproduce the observed lines at low metallicity end typically in the galactic outskirts. Specifically, for a given [N II], the model fails to reproduce a sufficient range of [O II] intensities to match the observed data. This limitation also provides an interpretation of the lack of model coverage in the *N2*–*N2O2* diagrams as presented above. Indeed, generation of certain emission lines in photoionization modeling is an intrinsically complicated mechanism. In addition to metallicity, the ionization parameter, and gas density, many other parameters, such as the shape and the intensity of ionizing radiation field, as well as the gas geometry and physics, play critical roles in predicting emission lines.

5.3. Impact of Rotation of Stars or Binary Populations

Spectral types of ionizing sources are a dominant factor for determining emission lines in photoionization processes and dependent on several assumptions in stellar population synthesis such as initial mass function, star formation history, stellar evolutionary tracks, stellar rotation, binary-population interactions, etc. It is mentioned in Levesque et al. [12] that the model is likely to require harder spectra to create sufficient ionizing fields and yield enough [S II] to successfully reproduce observed data. In this case, rotation of stars is suggested to achieve this. The stellar rotation is expected to create harder ultraviolet spectra for a longer period of time (Levesque et al. [57], Dorn-Wallenstein and Levesque [58]). Alternatively, binary stars are another possible requisite for maintaining hardness of ultraviolet radiation despite aging stellar populations (Stanway et al. [59]). Binary stellar population models make it possible to continuously reproduce observed data after 10 Myr in the BPT diagram and those at low metallicity or high redshifts in the [O III]-[S II] diagram (Xiao et al. [13], Zhang et al. [60]).

6. Summary

In this paper, we presented an investigation of spectral emission-line characteristics of H II regions in the nearby spiral galaxy NGC 2403. Spectroscopic observations were taken with the 2.16 m telescope at the National Astronomical Observatories of China, with supplementary data collected from the literature. As a consequence, a total of 188 H II regions in NGC 2403 were compiled in the final sample. A combined model of photoionization and stellar population synthesis was employed for comparing and interpreting the observed data.

Through the BPT diagnosis, nearly all of the samples were confirmed to originate from stellar irradiation and hence classified as star-forming regions. Radial profiles of metallicity for NGC 2403 were obtained by applying the four widely used strong-line diagnostics, *R23*, *O3N2*, *N2*, and *N2O2*, respectively, and all of them exhibited negative gradients, reflecting an inside-out evolution process for this galaxy, but the slopes and the dispersions of the gradients were different. *N2O2* resulted in the gradient with the steepest slope and the least dispersion, while the *N2* and *R23* indicators yielded the flattest gradients, and *R23* exhibited the largest dispersion.

Via a series of diagrammatic inspections on the four spectral indices *R23*, *O3N2*, *N2*, and *N2O2*, as well as individual emission lines within them, we demonstrated strong sensitivities of *O3N2* and *N2* to the ionization parameter; *R23* was comparatively less sensitive to the ionization parameter but had a double-value effect which was especially influential for the H II regions in NGC 2403; *N2O2* appeared to be the most robust diagnostic, in monotonic correlation with metallicity and with negligible sensitivity to the ionization parameter. By inspecting the individual emission lines, we disclosed that the [O III] emission line was strongly related with the age of ionizing stellar populations, in contrast to the [O II] or [N II] emission line in relatively weak relation with stellar population age. As a consequence, the indices *R23* and *O3N2* involving the [O III] emission line were dependent on stellar population age, while *N2* and *N2O2* were less affected by stellar population age, but the influence of the ionization parameter on *N2* became more serious for young stellar populations ($\lesssim$6 Myr). Thus, *N2O2* is suggested to more reliably indicate metallicity than the other three diagnostics. Furthermore, within the adopted photoionization model framework, the grid corresponding to an age of 0–1 Myr provided the best overall match to the observational data, whereas models older than 6.5 Myr showed poor agreement.

Author Contributions: Conceptualization, Q.-M.W. and Y.-W.M.; Methodology, Q.-M.W. and Y.-W.M.; Software, Q.-M.W.; Validation, Q.-M.W. and Y.-W.M.; Formal analysis, Q.-M.W.; Investigation, Q.-M.W. and Y.-W.M.; Resources, Y.-W.M. and L.L.; Data curation, Q.-M.W., Y.-W.M. and S.-T.W.; writing—original draft preparation, Q.-M.W. and Y.-W.M.; writing—review and editing, Q.-M.W., Y.-W.M., L.L., H.Z. and S.-T.W.; Visualization, Q.-M.W.; Supervision, Y.-W.M.; Project administration, Y.-W.M.; Funding acquisition, Y.-W.M., L.L. and H.Z. All authors have read and agreed to the published version of the manuscript.

Funding: This work is supported by the National Key R&D Program of China (No. 2022YFA1602902), the National Natural Science Foundation of China (NSFC, Nos. U2031106, U2031201, 12120101003, and 12373010), China Manned Space Project (Nos. CMS-CSST-2025-A06 and CMS-CSST-2021-B04), and the Youth Innovation Promotion Association CAS (id. 2022260).

Acknowledgments: We acknowledge the support of the staff of the Xinglong 2.16 m telescope. This work was partially supported by the Open Project Program of the Key Laboratory of Optical Astronomy, National Astronomical Observatories, Chinese Academy of Sciences. This research has made use of the NASA/IPAC Extragalactic Database (NED) which is funded by the National Aeronautics and Space Administration and operated by the California Institute of Technology; this research has also made use of the SAO/NASA's Astrophysics Data System Bibliographic Services.

Conflicts of Interest: The authors declare no conflicts of interest.

Appendix A. Index–Index Diagrams with Age Evolution

In this appendix, we present the index–index diagrams and the line–line diagrams, with the model grids at different ages superimposed, corresponding to the analysis in Section 4.3.2. The index–index diagrams are shown in Figures A1–A9, and the line–line diagrams are shown in Figures A10–A18.

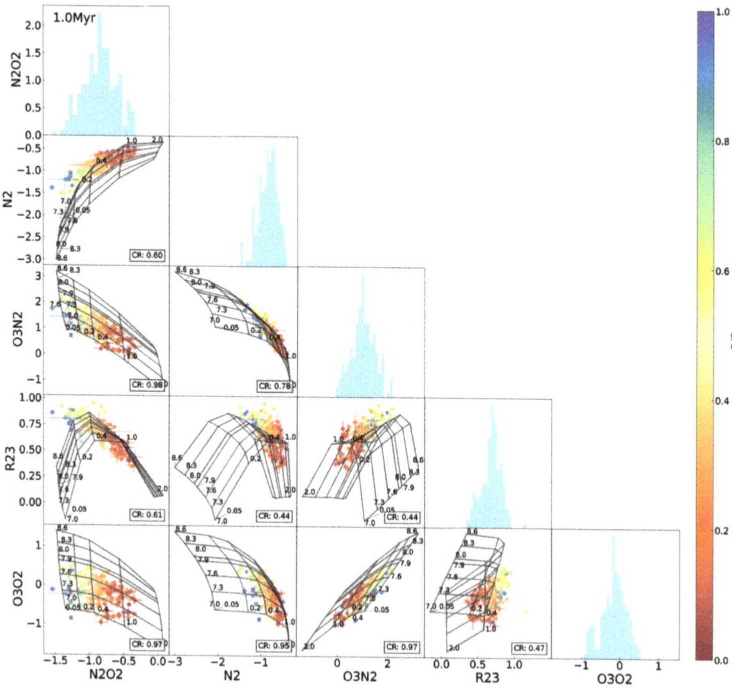

Figure A1. The same as Figure 5 but superimposing the 1.0 Myr model grids.

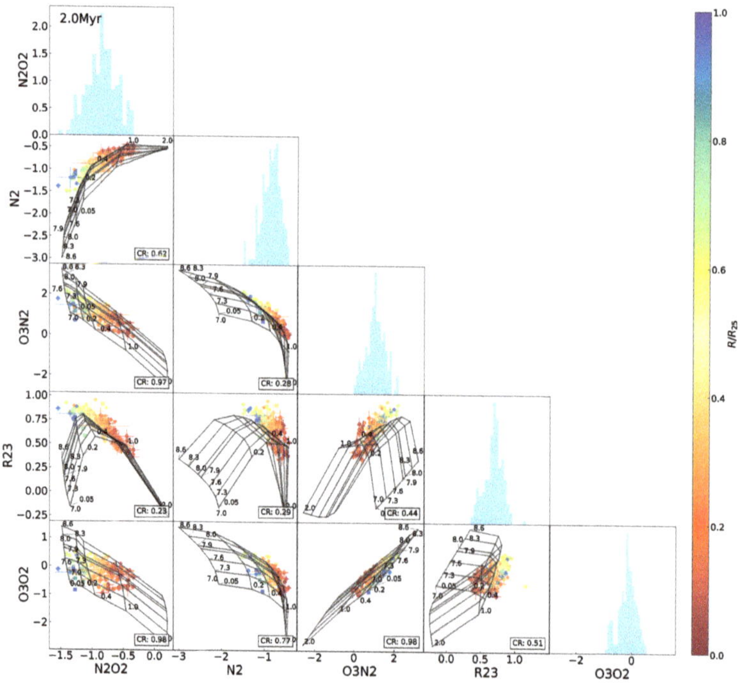

Figure A2. The same as Figure 5 but superimposing the 2.0 Myr model grids.

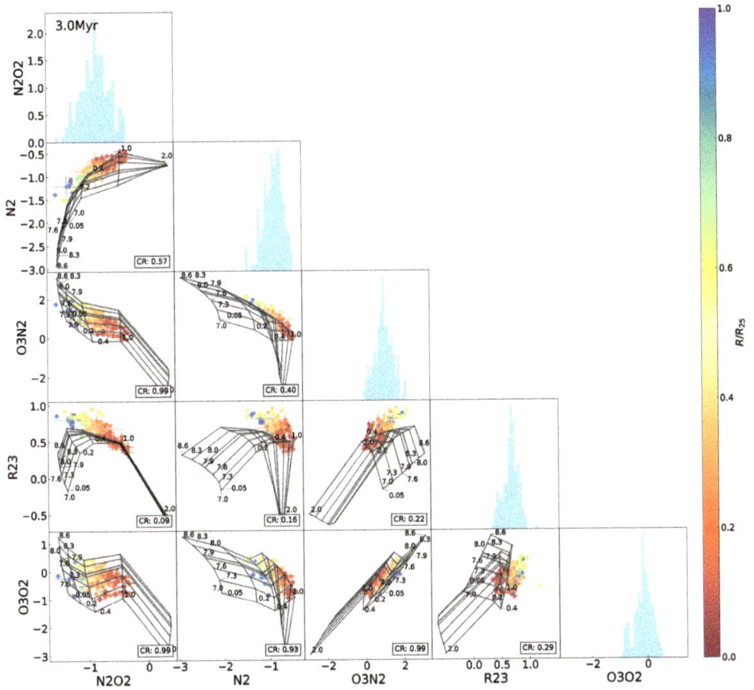

Figure A3. The same as Figure 5 but superimposing the 3.0 Myr model grids.

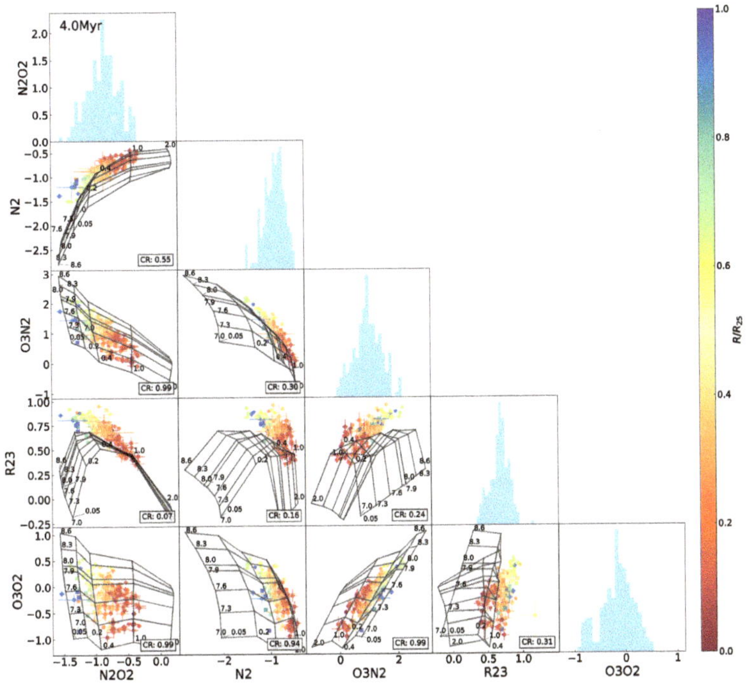

Figure A4. The same as Figure 5 but superimposing the 4.0 Myr model grids.

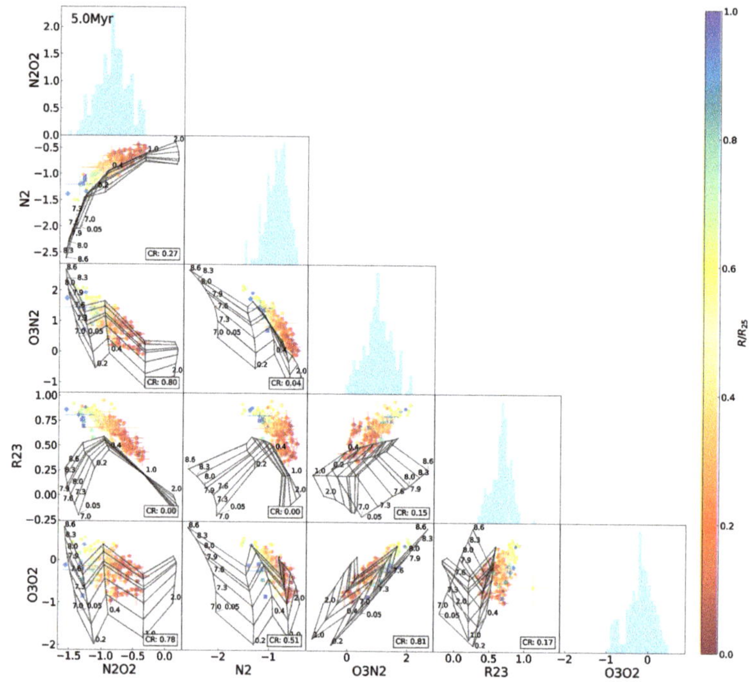

Figure A5. The same as Figure 5 but superimposing the 5.0 Myr model grids.

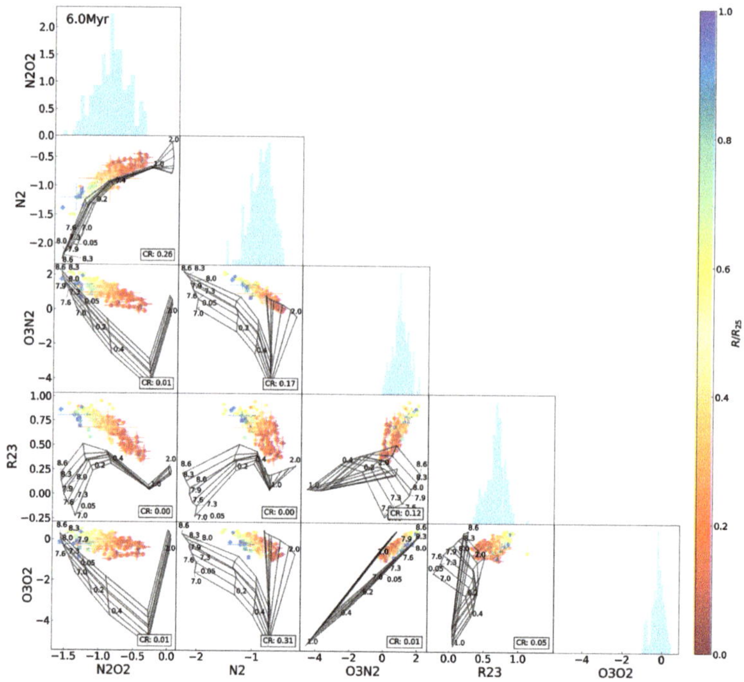

Figure A6. The same as Figure 5 but superimposing the 6.0 Myr model grids.

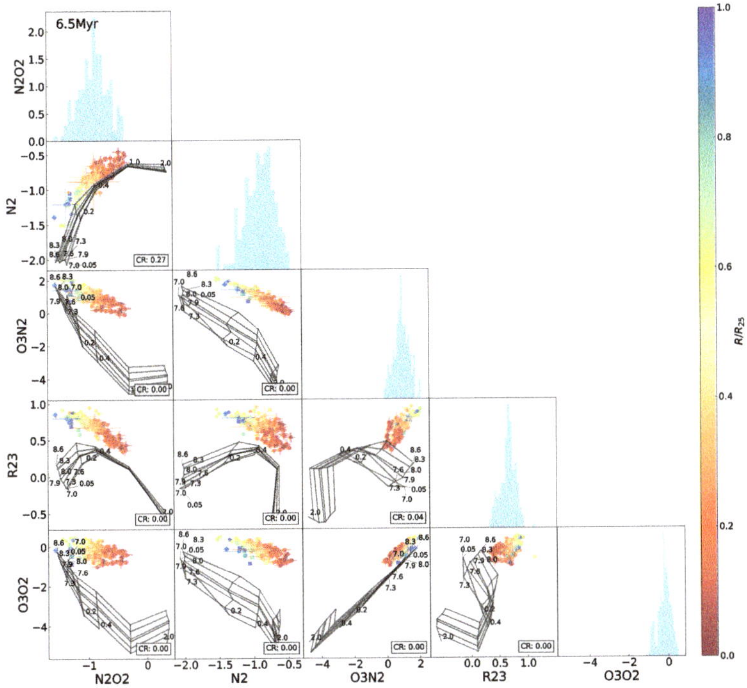

Figure A7. The same as Figure 5 but superimposing the 6.5 Myr model grids.

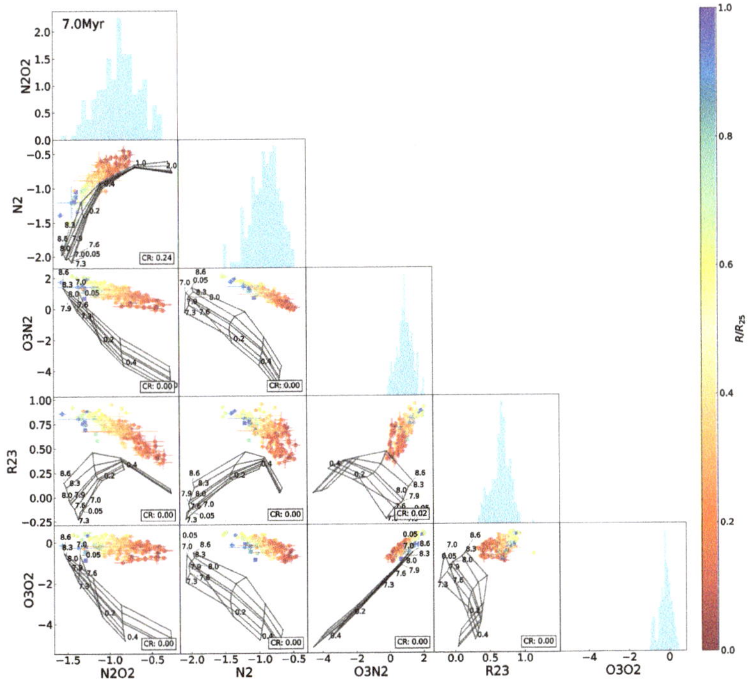

Figure A8. The same as Figure 5 but superimposing the 7.0 Myr model grids.

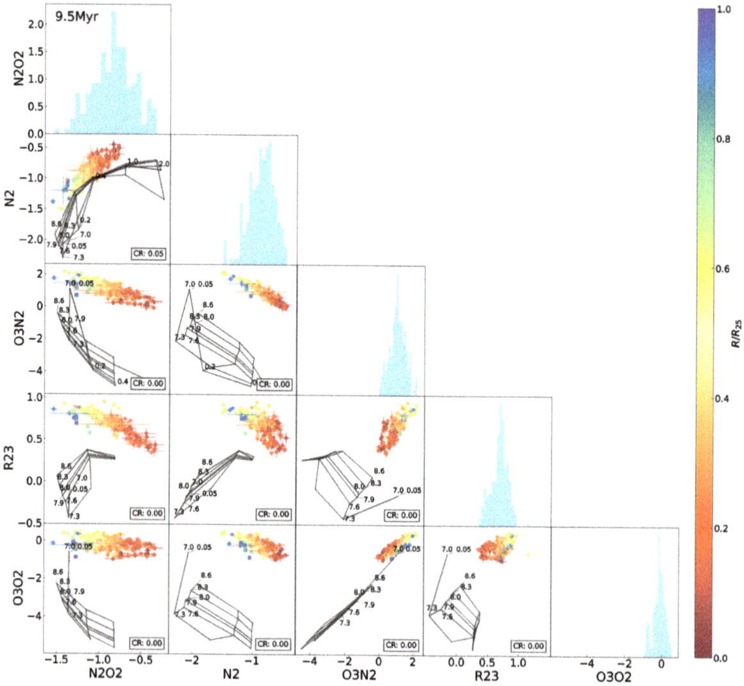

Figure A9. The same as Figure 5 but superimposing the 9.5 Myr model grids.

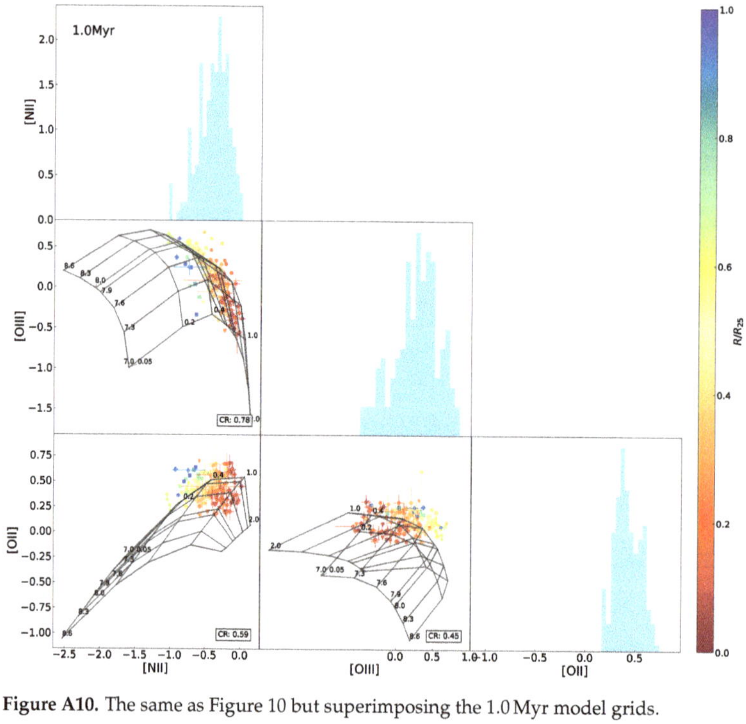

Figure A10. The same as Figure 10 but superimposing the 1.0 Myr model grids.

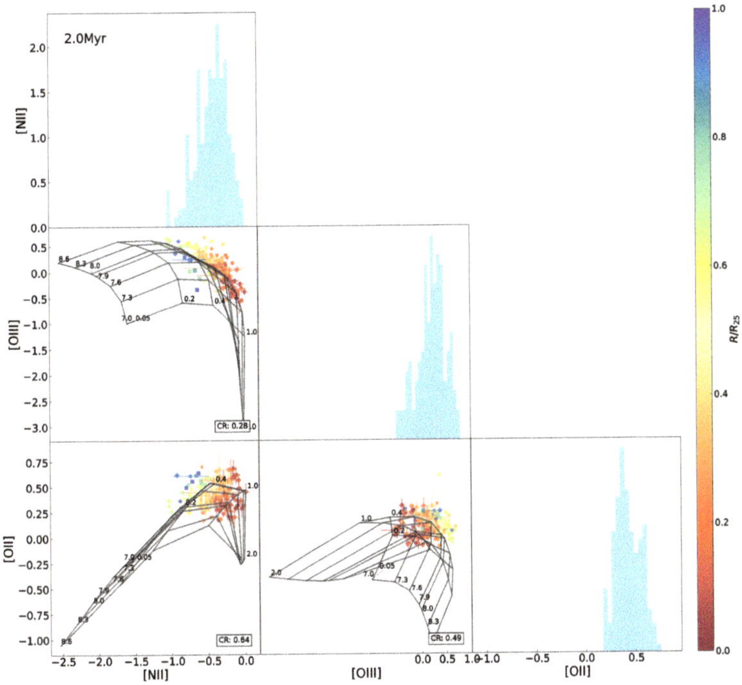

Figure A11. The same as Figure 10 but superimposing the 2.0 Myr model grids.

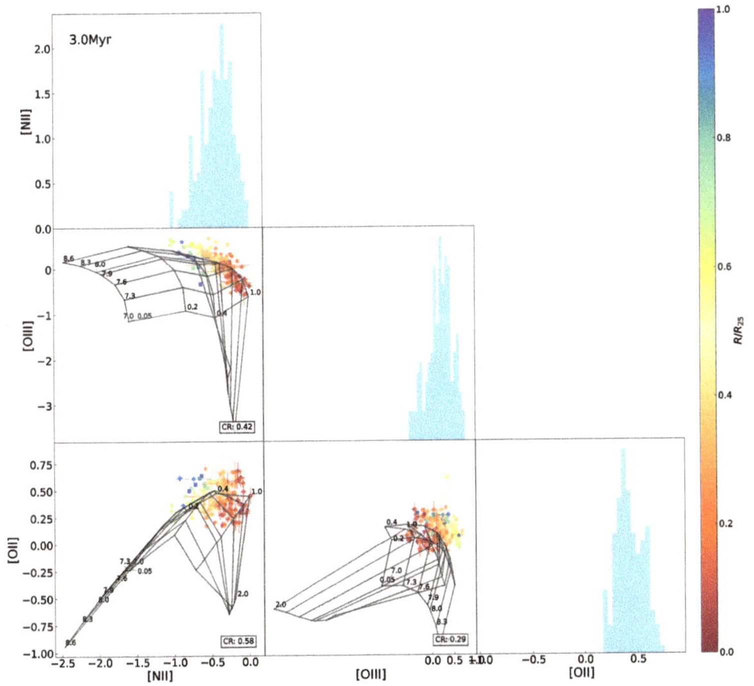

Figure A12. The same as Figure 10 but superimposing the 3.0 Myr model grids.

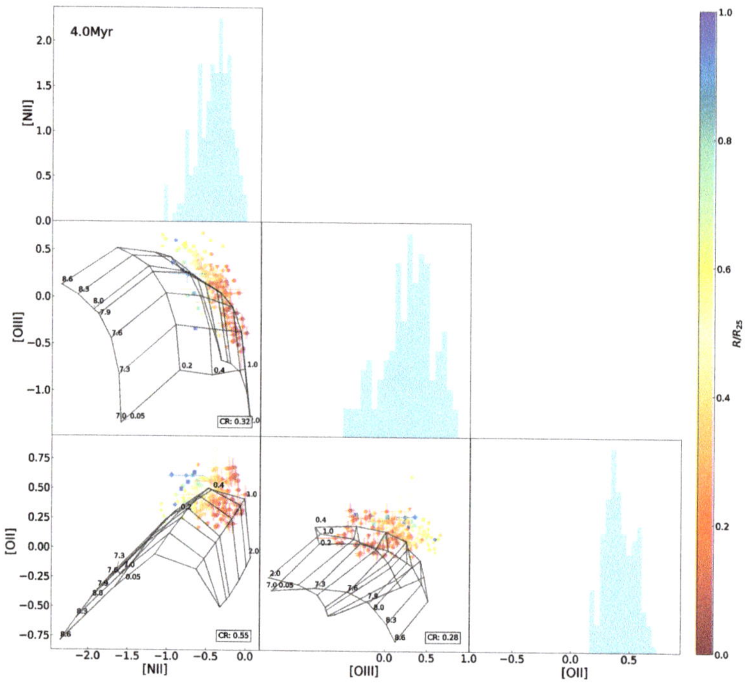

Figure A13. The same as Figure 10 but superimposing the 4.0 Myr model grids.

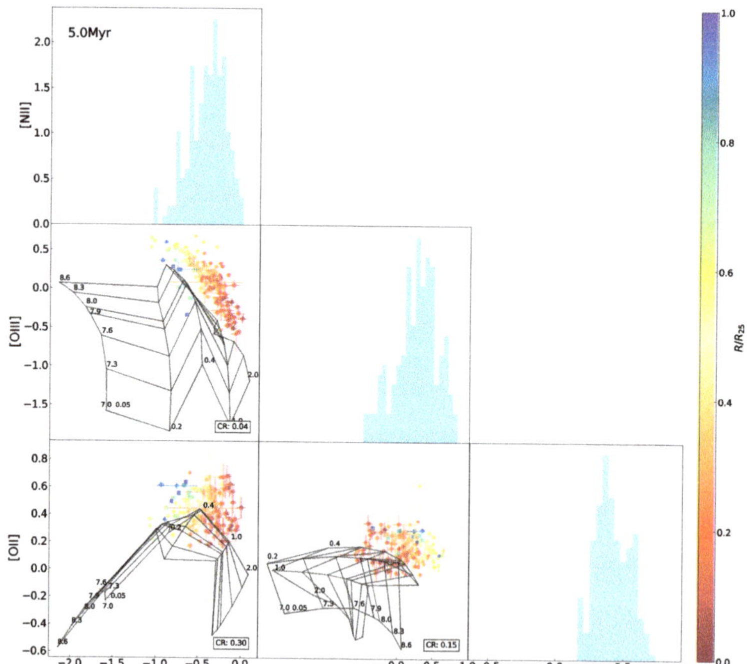

Figure A14. The same as Figure 10 but superimposing the 5.0 Myr model grids.

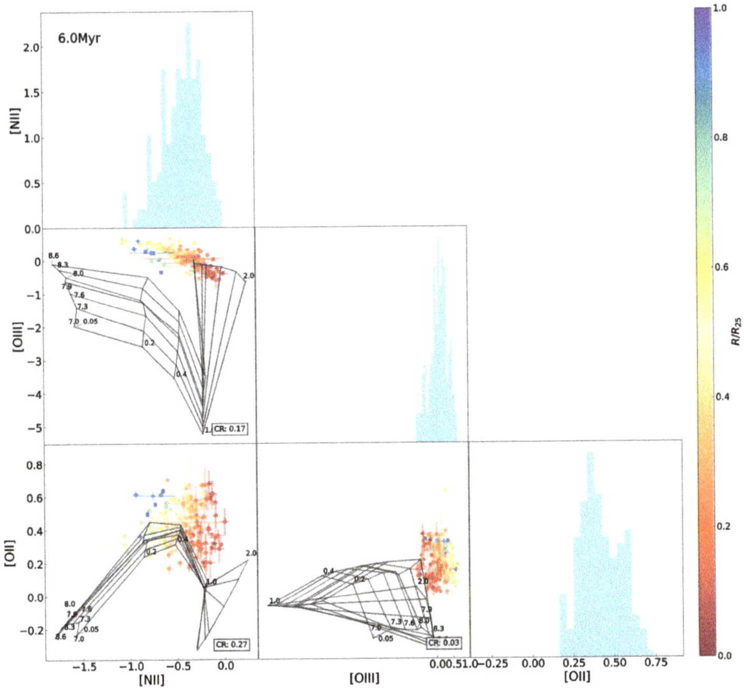

Figure A15. The same as Figure 10 but superimposing the 6.0 Myr model grids.

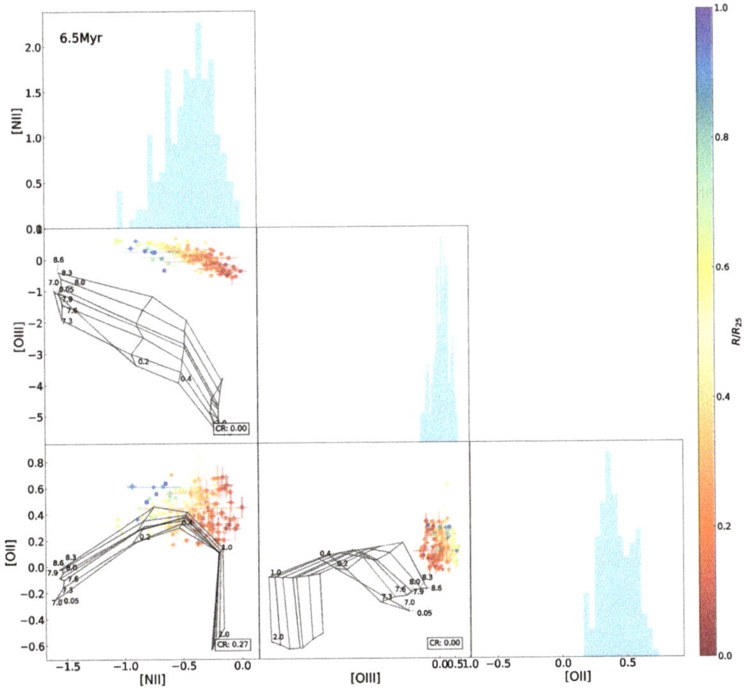

Figure A16. The same as Figure 10 but superimposing the 6.5 Myr model grids.

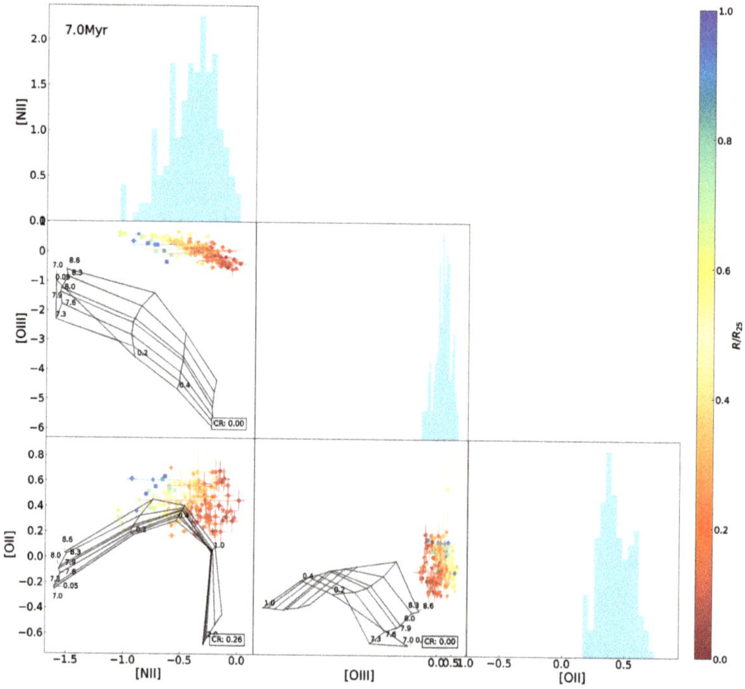

Figure A17. The same as Figure 10 but superimposing the 7.0 Myr model grids.

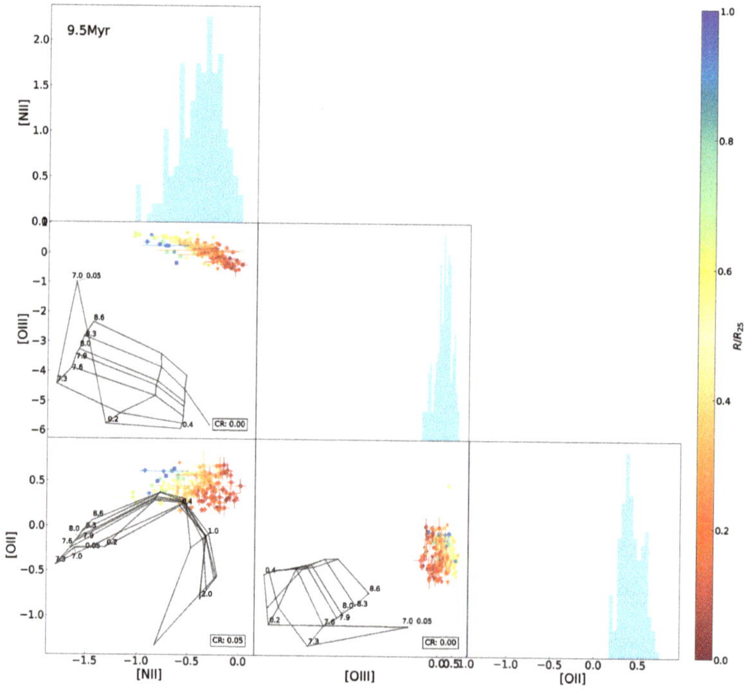

Figure A18. The same as Figure 10 but superimposing the 9.5 Myr model grids.

Notes

1 [O II]λ3727 is a sum of the double lines [O II]λ3726/[O II]λ3729, which are hardly resolvable for low-resolution spectroscopy and thus taken as one single line [O II]λ3727.

2 IRAF is distributed by the National Optical Astronomy Observatory, which is operated by the Association of Universities for Research in Astronomy, Inc., under a cooperative agreement with the National Science Foundation.

3 The *R*23 method has a double-value effect. We choose the higher branch in this work, because the H II regions in NGC 2403 are generally metal-rich according to our initial assessment.

4 In observational studies, a ratio between two emission lines for a certain ion at different ionization states is usually taken as a proxy of the ionization parameter (e.g., Kewley et al. [3], Mao et al. [18], Relaño et al. [53]). The emission-line indices [O III]$\lambda\lambda$4959, 5007/[O II]λ3727 (denoted as *O*3*O*2) and [S III]$\lambda\lambda$9069, 9531/[S II]$\lambda\lambda$6717, 6731 (denoted as *S*3*S*2 Kewley et al. [3], Kewley and Dopita [9], Morisset et al. [10], Dopita et al. [54], Garner et al. [55]) are the most widely used ionization tracers. Owing to the wavelength coverage in our observations, we employed *O*3*O*2 in the diagnoses.

5 Although the ionization parameter was taken into account, the sensitivity of the Pilyugin and Thuan [7] calibration to the ionization parameter was still found in our further analysis (Li et al., in preparation), which points to the complexity of the *R*23 index as a reliable or questionable estimator of metallicity.

6 https://massivestars.stsci.edu/starburst99, accessed on 8 August 2025.

7 In Figure 9, we note that some of the high-metallicity model points (in red and yellow) exhibit abrupt jumps at older ages. We suspect that these anomalies are artifacts arising from the model calculations. However, since this cannot be confirmed and they do not significantly impact our analysis, we chose to retain them.

References

1. Peimbert, M.; Peimbert, A.; Delgado-Inglada, G. Nebular Spectroscopy: A Guide on H ii Regions and Planetary Nebulae. *Publ. Astron. Soc. Pac.* **2017**, *129*, 082001. [CrossRef]
2. Maiolino, R.; Mannucci, F. De re metallica: The cosmic chemical evolution of galaxies. *Astron. Astrophys. Rev.* **2019**, *27*, 1–187. [CrossRef]
3. Kewley, L.J.; Nicholls, D.C.; Sutherland, R.S. Understanding Galaxy Evolution Through Emission Lines. *Annu. Rev. Astron. Astrophys.* **2019**, *57*, 511–570. [CrossRef]
4. Alloin, D.; Collin-Souffrin, S.; Joly, M.; Vigroux, L. Nitrogen and oxygen abundances in galaxies. *Astron. Astrophys.* **1979**, *78*, 200–216.
5. Pagel, B.E.J.; Edmunds, M.G.; Blackwell, D.E.; Chun, M.S.; Smith, G. On the composition of H II regions in southern galaxies–I. NGC 300 and 1365. *Mon. Not. R. Astron. Soc.* **1979**, *189*, 95–113. [CrossRef]
6. Pettini, M.; Pagel, B.E.J. [OIII]/[NII] as an abundance indicator at high redshift. *Mon. Not. R. Astron. Soc.* **2004**, *348*, L59–L63. [CrossRef]
7. Pilyugin, L.S.; Thuan, T.X. Oxygen Abundance Determination in H II Regions: The Strong Line Intensities-Abundance Calibration Revisited. *Astrophys. J.* **2005**, *631*, 231–243. [CrossRef]
8. Bresolin, F. The Oxygen Abundance in the Inner H II Regions of M101: Implications for the Calibration of Strong-Line Metallicity Indicators. *Astrophys. J.* **2007**, *656*, 186–197. [CrossRef]
9. Kewley, L.J.; Dopita, M.A. Using Strong Lines to Estimate Abundances in Extragalactic H II Regions and Starburst Galaxies. *Astrophys. J. Suppl. Ser.* **2002**, *142*, 35–52. [CrossRef]
10. Morisset, C.; Delgado-Inglada, G.; Sánchez, S.F.; Galbany, L.; García-Benito, R.; Husemann, B.; Marino, R.A.; Mast, D.; Roth, M.M. Photoionization models of the CALIFA H II regions. I. Hybrid models. *Astron. Astrophys.* **2016**, *594*, A37. [CrossRef]
11. Wang, W.; Liu, X.W.; Zhang, Y.; Barlow, M.J. A reexamination of electron density diagnostics for ionized gaseous nebulae. *Astron. Astrophys.* **2004**, *427*, 873–886. [CrossRef]
12. Levesque, E.M.; Kewley, L.J.; Larson, K.L. Theoretical Modeling of Star-Forming Galaxies. I. Emission-Line Diagnostic Grids for Local and Low-Metallicity Galaxies. *Astron. J.* **2010**, *139*, 712–727. [CrossRef]
13. Xiao, L.; Stanway, E.R.; Eldridge, J.J. Emission-line diagnostics of nearby H II regions including interacting binary populations. *Mon. Not. R. Astron. Soc.* **2018**, *477*, 904–934. [CrossRef]
14. Kong, X.; Lin, L.; Li, J.r.; Zhou, X.; Zhou, H.; Li, H.y.; Chen, F.z.; Du, W.; Fan, Z.; Mao, Y.w.; et al. Spectroscopic Observations of the Star Formation Regions in Nearby Galaxies. *Chin. Astron. Astrophys.* **2014**, *38*, 427–438. [CrossRef]
15. Tully, R.B.; Courtois, H.M.; Dolphin, A.E.; Fisher, J.R.; Héraudeau, P.; Jacobs, B.A.; Karachentsev, I.D.; Makarov, D.; Makarova, L.; Mitronova, S.; et al. Cosmicflows-2: The Data. *Astron. J.* **2013**, *146*, 86. [CrossRef]
16. Skrutskie, M.F.; Cutri, R.M.; Stiening, R.; Weinberg, M.D.; Schneider, S.; Carpenter, J.M.; Beichman, C.; Capps, R.; Chester, T.; Elias, J.; et al. The Two Micron All Sky Survey (2MASS). *Astron. J.* **2006**, *131*, 1163–1183. [CrossRef]

17. de Vaucouleurs, G.; de Vaucouleurs, A.; Herold, G.C., Jr.; Buta, R.J.; Paturel, G.; Fouque, P. *Third Reference Catalogue of Bright Galaxies*; Springer: New York, NY, USA, 1991.
18. Mao, Y.W.; Lin, L.; Kong, X. Internal Variations in Empirical Oxygen Abundances for Giant H II Regions in the Galaxy NGC 2403. *Astrophys. J.* **2018**, *853*, 151. [CrossRef]
19. Sivan, J.P.; Petit, H.; Comte, G.; Maucherat, A.J. Optical HII regions in NGC 2403. *Astron. Astrophys.* **1990**, *237*, 23.
20. Fan, Z.; Wang, H.; Jiang, X.; Wu, H.; Li, H.; Huang, Y.; Xu, D.; Hu, Z.; Zhu, Y.; Wang, J.; et al. The Xinglong 2.16-m Telescope: Current Instruments and Scientific Projects. *Publ. Astron. Soc. Pac.* **2016**, *128*, 115005. [CrossRef]
21. Corbin, T.E.; Warren, W.H., Jr. *International Reference Stars Catalog (Corbin 1991). Documentation for the Machine-Readable Version*; National Aeronautics and Space Administration (NASA): Greenbelt, MD, USA, 1991.
22. Filippenko, A.V. The importance of atmospheric differential refraction in spectrophotometry. *Publ. Astron. Soc. Pac.* **1982**, *94*, 715–721. [CrossRef]
23. Cardelli, J.A.; Clayton, G.C.; Mathis, J.S. The Relationship between Infrared, Optical, and Ultraviolet Extinction. *Astrophys. J.* **1989**, *345*, 245. [CrossRef]
24. Schlegel, D.J.; Finkbeiner, D.P.; Davis, M. Maps of Dust Infrared Emission for Use in Estimation of Reddening and Cosmic Microwave Background Radiation Foregrounds. *Astrophys. J.* **1998**, *500*, 525–553. [CrossRef]
25. Cid Fernandes, R.; Mateus, A.; Sodré, L.; Stasińska, G.; Gomes, J.M. Semi-empirical analysis of Sloan Digital Sky Survey galaxies—I. Spectral synthesis method. *Mon. Not. R. Astron. Soc.* **2005**, *358*, 363–378. [CrossRef]
26. Carnall, A.C. SpectRes: A Fast Spectral Resampling Tool in Python. *arXiv* **2017**, arXiv:1705.05165. [CrossRef]
27. Bruzual, G.; Charlot, S. Stellar population synthesis at the resolution of 2003. *Mon. Not. R. Astron. Soc.* **2003**, *344*, 1000–1028. [CrossRef]
28. Chabrier, G. Galactic Stellar and Substellar Initial Mass Function. *Publ. Astron. Soc. Pac.* **2003**, *115*, 763–795. [CrossRef]
29. Le Borgne, J.F.; Bruzual, G.; Pelló, R.; Lançon, A.; Rocca-Volmerange, B.; Sanahuja, B.; Schaerer, D.; Soubiran, C.; Vílchez-Gómez, R. STELIB: A library of stellar spectra at R ~ 2000. *Astron. Astrophys.* **2003**, *402*, 433–442. [CrossRef]
30. Gonzalez-Delgado, R.M.; Perez, E.; Tenorio-Tagle, G.; Vilchez, J.M.; Terlevich, E.; Terlevich, R.; Telles, E.; Rodriguez-Espinosa, J.M.; Mas-Hesse, M.; Garcia-Vargas, M.L.; et al. Violent Star Formation in NGC 2363. *Astrophys. J.* **1994**, *437*, 239. [CrossRef]
31. Calzetti, D. The Dust Opacity of Star-forming Galaxies. *Publ. Astron. Soc. Pac.* **2001**, *113*, 1449–1485. [CrossRef]
32. Osterbrock, D.E.; Ferland, G.J. *Astrophysics of Gaseous Nebulae and Active Galactic Nuclei*, 2nd ed.; University Science Books: Sausalito, CA, USA, 2006.
33. McCall, M.L.; Rybski, P.M.; Shields, G.A. The chemistry of galaxies. I. The nature of giant extragalactic H II regions. *Astrophys. J. Suppl. Ser.* **1985**, *57*, 1–62. [CrossRef]
34. Fierro, J.; Torres-Peimbert, S.; Peimbert, M. Chemical composition gradient in NGC 2403 and the stellar mass limit. *Publ. Astron. Soc. Pac.* **1986**, *98*, 1032–1040. [CrossRef]
35. Garnett, D.R.; Shields, G.A.; Skillman, E.D.; Sagan, S.P.; Dufour, R.J. Interstellar Abundance Gradients in NGC 2403: Comparison to M33. *Astrophys. J.* **1997**, *489*, 63–86. [CrossRef]
36. Bresolin, F.; Kennicutt, R.C., Jr.; Garnett, D.R. The Ionizing Stars of Extragalactic H II Regions. *Astrophys. J.* **1999**, *510*, 104–124. [CrossRef]
37. Garnett, D.R.; Shields, G.A.; Peimbert, M.; Torres-Peimbert, S.; Skillman, E.D.; Dufour, R.J.; Terlevich, E.; Terlevich, R.J. Carbon in Spiral Galaxies from Hubble Space Telescope Spectroscopy. *Astrophys. J.* **1999**, *513*, 168–179. [CrossRef]
38. van Zee, L.; Salzer, J.J.; Haynes, M.P.; O'Donoghue, A.A.; Balonek, T.J. Spectroscopy of Outlying H II Regions in Spiral Galaxies: Abundances and Radial Gradients. *Astron. J.* **1998**, *116*, 2805–2833. [CrossRef]
39. Berg, D.A.; Skillman, E.D.; Garnett, D.R.; Croxall, K.V.; Marble, A.R.; Smith, J.D.; Gordon, K.; Kennicutt, R.C., Jr. New Radial Abundance Gradients for NGC 628 and NGC 2403. *Astrophys. J.* **2013**, *775*, 128. [CrossRef]
40. Rogers, N.S.J.; Skillman, E.D.; Pogge, R.W.; Berg, D.A.; Moustakas, J.; Croxall, K.V.; Sun, J. CHAOS. VI. Direct Abundances in NGC 2403. *Astrophys. J.* **2021**, *915*, 21. [CrossRef]
41. Leitherer, C.; Schaerer, D.; Goldader, J.D.; Delgado, R.M.G.; Robert, C.; Kune, D.F.; de Mello, D.F.; Devost, D.; Heckman, T.M. Starburst99: Synthesis Models for Galaxies with Active Star Formation. *Astrophys. J. Suppl. Ser.* **1999**, *123*, 3–40. [CrossRef]
42. Vázquez, G.A.; Leitherer, C. Optimization of Starburst99 for Intermediate-Age and Old Stellar Populations. *Astrophys. J.* **2005**, *621*, 695–717. [CrossRef]
43. Sutherland, R.S.; Dopita, M.A. Cooling Functions for Low-Density Astrophysical Plasmas. *Astrophys. J. Suppl. Ser.* **1993**, *88*, 253. [CrossRef]
44. Allen, M.G.; Groves, B.A.; Dopita, M.A.; Sutherland, R.S.; Kewley, L.J. The MAPPINGS III Library of Fast Radiative Shock Models. *Astrophys. J. Suppl. Ser.* **2008**, *178*, 20–55. [CrossRef]
45. Schaller, G.; Schaerer, D.; Meynet, G.; Maeder, A. New Grids of Stellar Models from 0.8-SOLAR-MASS to 120-SOLAR-MASSES at Z=0.020 and Z=0.001. *Astron. Astrophys. Suppl. Ser.* **1992**, *96*, 269.

46. Schaerer, D.; Meynet, G.; Maeder, A.; Schaller, G. Grids of stellar models. II. From 0.8 to 120 M_solar at Z=0.008. *Astron. Astrophys. Suppl. Ser.* **1993**, *98*, 523.
47. Charbonnel, C.; Meynet, G.; Maeder, A.; Schaller, G.; Schaerer, D. Grids of stellar models. III. From 0.8 to 120 Msolar at Z = 0.004. *Astron. Astrophys. Suppl. Ser.* **1993**, *101*, 415.
48. Schaerer, D.; Charbonnel, C.; Meynet, G.; Maeder, A.; Schaller, G. Grids of stellar models. IV. From 0.8 to 120 M_solar at Z = 0.040. *Astron. Astrophys. Suppl. Ser.* **1993**, *102*, 339.
49. Baldwin, J.A.; Phillips, M.M.; Terlevich, R. Classification parameters for the emission-line spectra of extragalactic objects. *Publ. Astron. Soc. Pac.* **1981**, *93*, 5–19. [CrossRef]
50. Kewley, L.J.; Dopita, M.A.; Sutherland, R.S.; Heisler, C.A.; Trevena, J. Theoretical Modeling of Starburst Galaxies. *Astrophys. J.* **2001**, *556*, 121–140. [CrossRef]
51. Kauffmann, G.; Heckman, T.M.; Tremonti, C.; Brinchmann, J.; Charlot, S.; White, S.D.M.; Ridgway, S.E.; Brinkmann, J.; Fukugita, M.; Hall, P.B.; et al. The host galaxies of active galactic nuclei. *Mon. Not. R. Astron. Soc.* **2003**, *346*, 1055–1077. [CrossRef]
52. Kewley, L.J.; Ellison, S.L. Metallicity Calibrations and the Mass-Metallicity Relation for Star-forming Galaxies. *Astrophys. J.* **2008**, *681*, 1183–1204. [CrossRef]
53. Relaño, M.; Monreal-Ibero, A.; Vílchez, J.M.; Kennicutt, R.C. Spatially resolved study of the physical properties of the ionized gas in NGC 595. *Mon. Not. R. Astron. Soc.* **2010**, *402*, 1635–1648. [CrossRef]
54. Dopita, M.A.; Kewley, L.J.; Heisler, C.A.; Sutherland, R.S. A Theoretical Recalibration of the Extragalactic H II Region Sequence. *Astrophys. J.* **2000**, *542*, 224–234. [CrossRef]
55. Garner, R.; Kennicutt, R.; Rousseau-Nepton, L.; Olivier, G.M.; Fernández-Arenas, D.; Robert, C.; Martin, R.P.; Amram, P. NGC 628 in SIGNALS: Explaining the Abundance-ionization Correlation in H II Regions. *Astrophys. J.* **2025**, *978*, 70. [CrossRef]
56. Moustakas, J.; Kennicutt, R.C., Jr.; Tremonti, C.A.; Dale, D.A.; Smith, J.D.T.; Calzetti, D. Optical Spectroscopy and Nebular Oxygen Abundances of the Spitzer/SINGS Galaxies. *Astrophys. J. Suppl. Ser.* **2010**, *190*, 233–266. [CrossRef]
57. Levesque, E.M.; Leitherer, C.; Ekstrom, S.; Meynet, G.; Schaerer, D. The Effects of Stellar Rotation. I. Impact on the Ionizing Spectra and Integrated Properties of Stellar Populations. *Astrophys. J.* **2012**, *751*, 67. [CrossRef]
58. Dorn-Wallenstein, T.Z.; Levesque, E.M. A Comparison of Rotating and Binary Stellar Evolution Models: Effects on Massive Star Populations. *Astrophys. J.* **2020**, *896*, 164. [CrossRef]
59. Stanway, E.R.; Eldridge, J.J.; Becker, G.D. Stellar population effects on the inferred photon density at reionization. *Mon. Not. R. Astron. Soc.* **2016**, *456*, 485–499. [CrossRef]
60. Zhang, F.; Li, L.; Cheng, L.; Wang, L.; Kang, X.; Zhuang, Y.; Han, Z. Radiation fields of intermediate-age stellar populations with binaries as ionizing sources of H II regions. *Mon. Not. R. Astron. Soc.* **2015**, *447*, L21–L25. [CrossRef]

Disclaimer/Publisher's Note: The statements, opinions and data contained in all publications are solely those of the individual author(s) and contributor(s) and not of MDPI and/or the editor(s). MDPI and/or the editor(s) disclaim responsibility for any injury to people or property resulting from any ideas, methods, instructions or products referred to in the content.

universe MDPI

Review

A Comprehensive Guide to Interpretable AI-Powered Discoveries in Astronomy

Maggie Lieu

School of Physics & Astronomy, University of Nottingham, Nottingham NG7 2RD, UK;
maggie.lieu@nottingham.ac.uk

Abstract: The exponential growth of astronomical data necessitates the adoption of artificial intelligence (AI) and machine learning for timely and efficient scientific discovery. While AI techniques have achieved significant successes across diverse astronomical domains, their inherent complexity often obscures the reasoning behind their predictions, hindering scientific trust and verification. This review addresses the crucial need for interpretability in AI-powered astronomy. We survey key applications where AI is making significant impacts and review the foundational concepts of transparency, interpretability, and explainability. A comprehensive overview of various interpretable machine learning methods is presented, detailing their mechanisms, applications in astronomy, and associated challenges. Given that no single method offers a complete understanding, we emphasize the importance of employing a suite of techniques to build robust interpretations. We argue that prioritizing interpretability is essential for validating results, guarding against biases, understanding model limitations, and ultimately enhancing the scientific value of AI in astronomy. Building trustworthy AI through explainable methods is fundamental to advancing our understanding of the universe.

Keywords: machine learning; xAI; interpretable

1. Introduction

It is hard to believe that it was just 100 years ago, on the night of 5–6 October 1923, when Edwin Hubble, using the 100-inch Hooker telescope, discovered the first galaxy, besides our own galaxy the Milky Way—our galactic neighbor Andromeda [1]. This revelation expanded our perspective immeasurably, transforming the Milky Way from the entirety of existence to just one of countless galaxies scattered throughout the vast expanse of space. The revelation ignited a new era of astronomical exploration, with astronomers eager to map the universe and uncover its secrets. For centuries, the driving force behind many of the most profound discoveries has been the combination of advanced instrumentation and the discerning eye of human observers [2]. Even with advanced datasets, the manual scrutiny of scientists remains crucial, as evidenced by the recent identification of the Altieri Einstein ring within Euclid's archive [3]. Furthermore, dedicated amateur astronomers, armed with their own smaller instruments and a keen eye for detail, continue to play a vital role, contributing to discoveries including new asteroids or even exoplanets [4,5]. Even without telescope access, citizen science initiatives like Galaxy Zoo [4,6] have empowered the public to tackle datasets that were already beginning to challenge professional astronomers, classifying galaxies and discovering new phenomena. This collaborative human effort, aided by technology, has fueled a decade of remarkable discoveries, leading directly into the current era where the sheer scale of data presents unprecedented challenges.

The scale of modern astronomy has indeed grown exponentially. Early large-scale digital sky surveys, such as the Sloan Digital Sky Survey (SDSS) [7] and the Panoramic Survey Telescope and Rapid Response System (Pan-STARRS) [8], generated petabytes of data, creating massive archives. This trend is accelerating dramatically with next-generation facilities like the Vera C. Rubin Observatory conducting its Legacy Survey of Space and Time (LSST) [9], the Square Kilometre Array (SKA) [10], and space missions including Euclid [11], JWST [12], and the Roman Space Telescope [13]. Driven by significant advancements in detector sensitivity, resolution, and survey speed, these instruments are projected to generate data volumes measured not just in petabytes, but potentially exabytes over their lifetimes. Critically, this data deluge is not just voluminous but also characterized by immense complexity—multi-wavelength, time-sensitive, intricate structures. Their subtle signals are buried within noise and high dimensionality. Furthermore, the variety of data formats continues to expand, including vast mosaics of images, detailed spectra for millions of objects, high-cadence time series (light curves), complex integral field unit (IFU) datacubes mapping spatial and spectral information, and the outputs of increasingly sophisticated cosmological and astrophysical simulations (e.g., [14]). Handling such datasets is far beyond the capacity of traditional human analysis alone [15].

To extract meaningful scientific insights from this flood of information, artificial intelligence (AI), particularly machine learning (ML), has become an essential toolkit for modern astronomy, with usage growing exponentially over the last decade [16–18]. ML algorithms are uniquely suited for tasks intractable with traditional methods at these scales. This includes sophisticated pattern recognition in complex datasets, classification of diverse astronomical objects (e.g., stars, galaxies, supernovae types), regression for estimating physical parameters (e.g., photometric redshifts, stellar properties), anomaly detection to identify rare or unexpected phenomena, and dimensionality reduction to manage and interpret high-dimensional data spaces. Crucially, ML models can often identify subtle, non-linear correlations and patterns across vast numbers of features—relationships that might be imperceptible to human observers or beyond the scope of conventional statistical techniques [19].

This reliance on AI/ML stems directly from the inherent limitations of traditional methods in this data-rich era. Manual inspection cannot match the required speed and scalability. Human analysis is also susceptible to cognitive biases, perceptual limitations, and fatigue, potentially leading to inconsistencies in identifying and categorizing objects due to the subjective nature. Furthermore, many discoveries in astronomy come from unexpected anomalies where mistakes can become more likely, particularly when looking for rare or subtle signals. Conventional algorithms and human intuition struggle to effectively model or interpret the complex relationships within high-dimensional parameter spaces. They often rely on predefined criteria and may miss novel or rare events. AI/ML, in contrast, offers the potential for consistent analysis at scale, enabling the exploration of data complexity in ways previously impossible and can detect patterns beyond human perception.

The need for automation and speed is particularly acute in time-domain astronomy [20]. Facilities like the Zwicky Transient Facility [21,22] and, soon, the Rubin Observatory [9] generate millions of alerts nightly. Identifying and rapidly classifying interesting transient events such as supernovae (SNe) near peak brightness, gamma-ray bursts (GRBs), fast radio bursts (FRBs), or the elusive electromagnetic counterparts to gravitational wave (GW) events requires immediate attention within minutes or hours of detection [23]. This necessitates automated systems capable of real-time filtering, classification, and prioritization of these alerts to enable rapid follow-up observations with other telescopes. Human-in-the-loop processes are simply too slow for this discovery space and, thus, there is a compelling driver for AI adoption. Manually inspecting even a fraction of this is infeasible. Beyond transients, near-real-time applications are also emerging for tasks like optimizing telescope scheduling based on observing conditions and scientific priorities, or performing rapid data quality assessments [24].

While the power and success of AI in tackling these challenges are undeniable, leading to numerous discoveries (as this review will detail), the increasing sophistication of the ML models employed often comes at the cost of transparency. Many powerful algorithms, particularly deep learning models excelling at the tasks described above, function as "black boxes", where the internal logic behind a specific prediction or classification is not readily apparent [25]. In scientific discovery, however, understanding why a model arrives at a conclusion is as important as the conclusion itself. Trust, validation against known physics, reliability, and the ability to identify whether a discovery is genuine or merely an artifact of the model or data are paramount [26]. This critical need for understanding and trustworthiness motivates the growing field of interpretable machine learning and explainable AI (xAI) within astronomy. Furthermore, developing AI systems capable of more explicit reasoning may offer pathways to both more robust and more understandable models.

Therefore, this review proceeds by first establishing the necessary context for trustworthy AI. We will begin by exploring the foundational concepts of xAI and transparency, outlining their critical importance for validation and reliability in astronomical research (Section 2). Subsequently, we survey key astronomical domains where AI has enabled significant discoveries, examining the methods used and highlighting where interpretability considerations arise (Section 3). Following this is a comprehensive guide to specific emerging xAI techniques and reasoning models (Section 4), and we end by discussing open challenges, promising trends, and the collaborative efforts needed to advance scientifically valuable AI (Section 5).

2. Foundations of Trustworthy AI in Astronomy

The indispensable role of AI and machine learning (ML) in handling complex modern astronomical data brings critical considerations regarding trustworthiness and comprehensibility. Many powerful ML models, especially deep neural networks adept at image analysis and pattern recognition, often operate as "black boxes" [25]. Their internal mechanics, involving potentially billions of parameters tuned through complex optimization, lack inherent transparency. This opacity hinders the understanding of precisely how outputs like classifications or anomaly scores are generated, moving beyond simple input–output checks.

Trustworthiness in AI extends beyond model complexity. As data-driven systems, ML performance hinges on training data quality and representativeness. Astronomical datasets, often heterogeneous, incomplete, or affected by instrumental systematics, can lead models to inadvertently exploit irrelevant features or biases. Iterative model development cycles (feature selection, tuning, evaluation) compound these issues. Without rigorous validation and understanding of model behavior, this can lead to overfitting, learning noise or spurious correlations, or the production of results that seem desirable but are ungeneralizable or scientifically unsound. Furthermore, the vulnerability of vision models to subtle perturbations highlights robustness challenges [27]. Lack of transparency exacerbates these risks, complicating independent verification.

Therefore, in astronomy, where empirical evidence, validation against physical principles, and reproducibility are foundational, high predictive accuracy alone is insufficient. Understanding the reasoning behind AI results is crucial for building trust, verifying findings, fostering new insights, and enabling effective human–machine collaboration [26]. This necessitates examining the core concepts underpinning trustworthy AI: transparency, interpretability, and explainability. While often used interchangeably, they represent distinct facets of understanding AI systems:

- Transparency refers to the accessibility and understandability of the model's internal mechanics. This involves knowledge of the architecture, algorithms, learned parameters (e.g., weights in a neural network), and potentially the training data [25,28]. A model can be transparent (e.g., open source and well-documented) yet still lack intuitive comprehensibility for domain experts.
- Interpretability captures the extent to which a human—especially a domain expert—can understand the relationship between inputs and outputs, or the decision logic of a model [29,30]. Models like linear regression or decision trees are inherently interpretable; their outputs can be directly tied to feature contributions. In contrast, deep neural networks and ensemble models (e.g., random forests) present significant challenges to interpretation due to their layered complexity and non-linearity.
- Explainability (xAI) refers to methods that provide human-understandable post hoc explanations for model behavior, particularly for opaque models [31,32]. These explanations may be local (explaining a single decision) or global (overall behavior). Tools like SHAP [33], LIME [34], and attention-based visualizations [35] approximate model reasoning or highlight influential features, providing insights without full interpretability.

However, post hoc xAI methods have limitations. They often approximate reasoning and can be incomplete or misleading, particularly when justifying decisions retroactively [36,37]. Critics argue that this reliance might legitimize black-box use where interpretable alternatives exist, spurring a movement towards models interpretable by design [38]. Interpretability holds particularly high stakes in astronomy. Complex AI risks learning from instrumental artifacts, biases, or spurious correlations, potentially leading to unsound conclusions [39]. Interpretability allows astronomers to verify if models focus on astrophysically meaningful features (e.g., galaxy morphology, spectral lines) rather than confounders. Explainable models can also drive discovery by highlighting subtle, unnoticed patterns for further investigation. Explainability enhances scientific accountability. When unexpected results arise (e.g., a new transient type), robust explanations build community confidence and guide follow-up experiments. Understanding failure modes (e.g., poor performance in specific data subsets) is crucial for improving pipelines and model design. Furthermore, in an era of automated surveys, explainability is vital for ethical and equitable science. Historical data biases (e.g., from sky coverage limits, human classification) can be unintentionally amplified by AI. Interpretable methods are, thus, essential for identifying, quantifying, and mitigating these biases, promoting more inclusive and robust outcomes. Regardless of the approach, pursuing transparency, interpretability, and explainability is fundamental to the scientific value and reliability of AI in astronomy. Integrating these qualities into AI development, evaluation, and deployment ensures that derived insights are credible, reproducible, and advance our understanding of the universe. Grasping these concepts and the surrounding debate provides a necessary framework for assessing AI applications in astronomy.

3. AI Applications and Discoveries in Astronomy

Having established the foundational concepts of trustworthy AI (Section 2), this section surveys the diverse applications and significant discoveries enabled by AI and machine learning across various astronomical domains. The aim is to illustrate the transformative impact of these techniques while also highlighting, where relevant, the nature of the models employed and the associated considerations for trustworthiness.

3.1. Strong Lensing

Strong gravitational lensing provides a unique window into the Universe. These rare events occur when a massive foreground object (e.g., galaxy, cluster) bends and magnifies light from a distant background source, creating multiple images, arcs, or complete Einstein rings [40]. Studying these systems is invaluable for mapping dark matter, probing lens structure, measuring cosmological parameters like the Hubble constant [41], and observing otherwise faint distant sources, enabling glimpses into the early Universe and tests of general relativity [42].

First discovered in 1979 [43], strong lenses were identified for decades primarily through visual inspection or targeted searches, yielding several hundred confirmations. However, the advent of large-scale surveys and machine learning (ML) has dramatically increased the known sample. An early ML success was demonstrated in the Strong Lens Finding Challenge [44], where the winning support vector machine (SVM) [45], an algorithm finding optimal separating hyperplanes, outperformed human inspectors on simulated images by detecting subtle statistical differences. This SVM notably relied on hand-crafted features based on prior knowledge (e.g., arc morphology), offering some interpretability as the importance of specific, physically motivated features could be assessed. While this reliance potentially limited discovering novel configurations, the authors noted that it could reduce overfitting compared to pixel-based methods, despite cautioning about potential biases. Applying this SVM to the Kilo Degree Survey (KiDS) [46] revealed new candidates but also highlighted biases necessitating human intervention, particularly the model's tendency to misclassify non-lens objects like spiral galaxies as strong lenses due to the underrepresentation of these contaminating object types in the training data, necessitating human intervention, a process unscalable for future datasets [47].

This scalability challenge spurred the adoption of convolutional neural networks (CNNs), which automatically learn hierarchical features from raw pixels (e.g., [48,49]). CNNs excel at image recognition, and recent applications to DESI Legacy Imaging Surveys [50] and Euclid Quick Release data [51] have yielded thousands of new candidates, vastly expanding the known sample (previously <1000 confirmed). Despite their performance, the complex, multi-layered structure of CNNs makes them archetypal "black boxes", lacking the inherent transparency of feature-based SVMs. Understanding their decision process requires post hoc explainable AI techniques. Applying such techniques is crucial for validating candidates, distinguishing real lenses from mimics or artifacts, and building trust, especially as surveys like Euclid anticipate over 100,000 candidates, exceeding human vetting capacity. Bridging this performance–understanding gap through interpretation, particularly for rare lenses, is an area of emerging research (e.g., [52]).

3.2. Galaxy Morphology

Just as in lens finding, AI has also revolutionized the study of galaxy morphology, another area where visual classification has been central. Understanding galaxy formation and evolution critically depends on robust morphological classification—distinguishing spirals, ellipticals, irregulars, and mergers. These morphologies offer insights into dynamical histories, stellar populations, and environmental influences. Modern imaging surveys (SDSS, Pan-STARRS, DESI, Euclid, LSST) produce vast datasets of millions to billions of galaxy images, necessitating automated, scalable classification approaches.

Early efforts like the Galaxy Zoo project [6] successfully crowdsourced morphological classifications, yielding high-quality labeled datasets and catalyzing ML adoption in astronomy. Traditional ML models (e.g., random forests, SVMs) trained on manually extracted features (concentration, asymmetry, Gini coefficient) showed promise (e.g., [53,54]).

Their key advantage was inherent interpretability via feature importance rankings, helping astronomers understand which measurable properties drove classification.

Deep learning, particularly CNNs, transformed morphological classification, removing the need for hand-crafted features. Studies by Dieleman et al. [55] and Domínguez Sánchez et al. [56] demonstrated CNNs outperforming traditional methods, especially with large labeled datasets. Consequently, CNNs have generated large-scale morphological catalogs crucial for galaxy evolution studies.

However, this improved performance comes at the cost of interpretability. CNNs are highly accurate, yet their internal decision making based on visual features remains unclear. This opacity limits scientific trust and potential discovery. Are CNNs identifying canonical features (spiral arms, bars, bulges, tidal tails) or overfitting to artifacts or spurious correlations?

To probe the inner workings of these CNNs, techniques like saliency maps, Grad-CAM, and attention mechanisms (Section 4) visualize which image regions most influence classification. For instance, Bhambra et al. [57] used such tools to assess if CNNs focused on expected features like bars or merger tails. These visualizations provide qualitative insights and act as sanity checks, verifying alignment with meaningful astrophysical features or revealing unexpected patterns warranting further investigation.

3.3. Transient Detection and Classification

Moving from static morphology to dynamic phenomena, transient astronomy studies events evolving on short timescales (milliseconds to years) before fading or changing significantly. Unlike stars or galaxies, transients often signal cataclysmic events like supernovae (SNe), gamma-ray bursts (GRBs), fast radio bursts (FRBs), tidal disruption events (TDEs), or various variable stars. Modern surveys (e.g., ZTF, Gaia, and, soon, LSST) generate immense data streams—thousands to millions of alerts nightly—creating a data-rich but time-poor field [9]. This sheer volume and diverse mix of known variables, artifacts, and potential novel events demand sophisticated automation for rapid identification, classification, and follow-up.

Machine learning provides essential tools for this challenge. Its primary contribution is scalability, enabling crucial real-time filtering of vast alert streams to distinguish interesting signals from non-astrophysical detections or common variables, a task impossible manually [58]. Furthermore, transients involve complex, multi-modal data (light curves, spectra, host properties, images). ML, especially deep learning, is effective at automatically extracting features and identifying intricate patterns across these data types, often surpassing traditional algorithms [59]. ML classifiers can also provide vital probabilistic outputs for incomplete or ambiguous data [60], guiding follow-up decisions under uncertainty [61]. Crucially, ML powers anomaly detection algorithms, sifting through millions of events to flag outliers deviating from known classes, paving the way for discovering new astrophysical phenomena.

However, the effective deployment and scientific acceptance of these techniques critically depend on understanding and trusting their outputs. This brings interpretability and explainability to the forefront. Scientific trust and validation are paramount. Astronomers need confidence that an ML classification (e.g., SN Ia vs. SN II) or anomaly flag relies on relevant astrophysical characteristics (light curve shape, color evolution) rather than spurious correlations or artifacts [62]. This understanding directly impacts rapid follow-up prioritization, as interpretable justifications boost confidence when allocating scarce telescope time for confirmation, often needed within hours [17].

Moreover, interpretability is intrinsically linked to discovery via anomaly detection. An ML model merely flagging an event as anomalous offers limited insight. Explainable methods highlighting why an event is unusual (specific features or behaviors) are essential for characterizing potential novel phenomena and understanding the underlying physics [59]. For instance, Zhang et al. [63] demonstrated that saliency maps (Section 4.2), which highlight regions of input data most influential to a model's output, can not only enhance transient features, even without extensive pre-processing like interference dedispersion, but their analysis also led to the uncovering of a new pulsar in archival data, showcasing how interpretability can directly contribute to discovery. Finally, interpretability fosters effective human–AI collaboration. Explanations (feature importance, attention maps, prototype comparisons) allow domain experts to synergize their knowledge with ML predictions, enabling faster validation, building intuition, and accelerating discovery. Thus, while ML provides the indispensable engine for transient data, xAI provides the crucial layer of understanding and verification needed to ensure the reliability and scientific value of AI-driven transient astronomy.

3.4. *Galaxy Cluster Mass Estimation*

Beyond the high-volume, high-speed challenges of transient astronomy, the need for interpretability also arises in specific astrophysical modeling tasks. Galaxy clusters serve as a prime example. As the largest known gravitationally bound systems, containing hundreds to thousands of galaxies within massive dark matter halos (10^{14}–$10^{15} M_{\odot}$), accurately estimating their masses provide independent measurements for cosmological parameters like Ω_m, σ_8, and w through cluster number counts aids in breaking degeneracies [64]. However, this is challenging using observational data like X-ray emission tracing the hot intracluster medium. Low photon counts can affect distant and low mass clusters leading to statistical uncertainties, and the complex physics of the intracluster medium, unlike the simpler Cosmic Microwave Background, introduces systematic uncertainties such as hydrostatic bias, even when photon statistics are good.

Addressing this challenge, Ntampaka et al. [65] applied deep learning to estimate galaxy cluster masses directly from simulated, low-resolution, single-color mock images mimicking observations from the Chandra X-ray Observatory. They employed a relatively simple CNN architecture, for which, despite the architectural simplicity compared to some modern networks, the resulting mass estimates yielded mass estimates with small biases relative to the simulation's true masses.

However, recognizing the black box nature of even simpler CNNs, a key part of their work involved visually interpreting the trained model. They adapted a gradient ascent approach inspired by techniques like Google's DeepDream [66] to modify input images, identifying features that maximally activated neurons or drove higher mass predictions. This method served as a post hoc explainability technique aimed at feature attribution through visualization.

Their analysis revealed that the CNN was most sensitive to X-ray photons in the cluster outskirts, not the brighter inner regions. This was significant as it aligned with independent analyses indicating outskirts contain substantial mass information. Ntampaka et al. [65] concluded that the work highlights the utility of interpreting ML models. The visualization provided plausible physical reasoning for the model's predictions, increasing trust beyond simple accuracy metrics. It showcased the value of applying explainability methods, even early ones, to validate AI against physical intuition and ensure models leverage meaningful signals within complex astronomical data.

3.5. Galactic Archaeology

Mapping the structural components of the Milky Way, its disk, halo, and stellar populations, is central to understanding its formation and evolutionary history, a field known as galactic archaeology. Large-scale astrometric surveys, with Gaia [67] being a prominent example, provide unprecedented high-dimensional datasets containing stellar positions, motions, and magnitudes for over a billion stars that have fueled numerous discoveries. For instance, machine learning has been pivotal in identifying rare objects like hypervelocity stars (HVSs). Ref. [68] used an artificial neural network on Gaia DR1 data to identify thousands of candidates, which, after refinement, quadrupled the number of known HVSs. Neural networks have also extended the known memberships of hundreds of open clusters down to Gaia's detection limit [69]. However, extracting coherent structures from this vast data space often relies on unsupervised machine learning techniques [70], especially clustering algorithms, which identify intrinsic groupings without predefined labels. These methods have been instrumental in uncovering new stellar streams [71] and substructures within the halo and disk [72]. In this context, interpretability focuses less on model predictions and more on understanding the physical nature and significance of the emergent structures, posing unique challenges in validation and scientific inference. Unlike supervised learning, where ground truth provides a benchmark, clustering in galactic archaeology requires domain-informed interpretation and cross-validation with simulations or follow-up observations to confirm astrophysical relevance. Prototype-based explanations (Section 4.5), where a cluster's centroid acts as a representative prototype, offer one avenue for interpreting these clustering results.

4. Interpretable Machine Learning Methods

As machine learning becomes deeply embedded in the astronomical discovery pipeline, the imperative for interpretability grows more urgent. In a domain where scientific validation, reproducibility, and alignment with physical principles are essential, models must not only perform well but also be understood and trusted by human users.

Interpretable machine learning (IML) provides tools and frameworks for unpacking the decision-making processes of AI models. These methods offer insights into model behavior through different lenses: feature importance highlights which inputs matter most; attention mechanisms and saliency maps suggest where the model is focusing; rule-based models articulate decision logic; and post hoc tools like SHAP and LIME help explain specific predictions. These techniques can validate outputs, identify biases, uncover errors, and even inspire new scientific hypotheses.

Several IML approaches have already been adapted to astronomy, including symbolic regression for deriving physically meaningful relationships and hybrid strategies that combine spatial and feature-level insights, for example, integrating attention maps with SHAP values to understand both what and where a model is attending. Given the complexity of astronomical data, including high-dimensional photometry, time-domain variability, and multi-wavelength imaging, no single interpretability technique is sufficient. Hybrid methods are often needed to capture the full range of insights.

However, each technique comes with limitations. Feature importance methods can struggle with correlated inputs and high-dimensional spaces. Attention maps do not always align with human intuition, and saliency methods may be sensitive to noise or adversarial perturbations. SHAP and LIME, while flexible and model-agnostic, can be computationally intensive and sometimes unfaithful to the model's true reasoning, especially in extrapolative settings. Faithfulness, in the context of explainable AI, refers to how accurately an explanation reflects the model's underlying reasoning process for a specific prediction or behavior. A faithful explanation should genuinely represent why the model made a particular decision, rather than being a plausible but potentially misleading justification. Rule-based models, though transparent, often lack the performance required for tasks involving subtle or complex patterns.

Interpretability techniques can also be broadly classified as model-specific or model-agnostic. Model-specific methods, such as gradients or attention mechanisms in CNNs, leverage internal model structures to provide efficient, architecture aware insights. Model-agnostic techniques like SHAP and LIME offer broader applicability and support comparison across model types, though often at the expense of faithfulness and computational efficiency. A growing body of research suggests that combining both can enhance understanding without compromising performance.

Quantifying interpretability remains an open challenge. Metrics such as fidelity (alignment with the model's true logic), stability (robustness to input perturbations), simplicity (e.g., rule length or tree depth), and comprehensibility (human interpretability) are commonly used. In astronomy, domain-specific criteria such as consistency with known physics and localization accuracy are also increasingly applied to evaluate explanation quality.

Taken together, IML techniques offer astronomers a powerful toolkit for evaluating, debugging, and refining machine learning models. They offer distinct pathways to understanding how AI systems process data and arrive at predictions which we discuss in this section. Table 1 provides a quick overview of the use cases for each data type.

Table 1. Guide to selecting interpretability methods based on data type and goal.

Interpretability Goal	**Image Data**	**Tabular Data**	**Time Series Data**
What matters? (global)	SHAP PI	SHAP PI/GI	SHAP PI
Why this decision? (local)	Saliency LIME/SHAP attention -	- LIME/SHAP - rule-based	Saliency LIME/SHAP attention -
Where is it looking?	Saliency LIME/SHAP attention	- - -	Saliency - attention
How does it generalise?	Symbolic regression	Symbolic regression	Symbolic regression
What is similar?	Prototype/ Exempler	Prototype/ Exempler	Prototype/ Exempler

4.1. Feature Importance

A fundamental step towards understanding machine learning models, particularly those trained on tabular data, involves identifying which input features most significantly influence their predictions. Feature importance techniques quantify the contribution of each input variable to the model's performance or decision-making process. These methods are valuable across various astronomical applications, such as predicting stellar parameters from photometry or spectra, classifying supernova types based on light curve characteristics, or identifying galaxy clusters from survey data, as understanding the role of individual observational variables (e.g., specific colors, spectral line ratios, morphological parameters) is crucial for scientific validation and insight. Two commonly employed techniques are Gini importance (intrinsic to tree-based models) and permutation importance (model-agnostic).

4.1.1. Gini Importance

Gini importance (GI), also known as mean decrease in impurity (MDI), assesses the importance of features in tree-based ensemble models like random forests and gradient boosted trees. Decision trees build partitions by recursively splitting nodes based on the feature and threshold that yield the greatest reduction in node impurity. The Gini importance quantifies how much the impurity decreases when a specific feature is used for splitting data at each node. Common impurity measures are the Gini impurity (for classification) or variance (for regression). The Gini impurity for a node m with K classes is calculated as

$$I_G(m) = \sum_{k=1}^{K} p_{mk}(1 - p_{mk}) = 1 - \sum_{k=1}^{K} p_{mk}^2 \quad (1)$$

where p_{mk} is the proportion of samples of class k in node m.

The importance of a feature j at a specific node m is the impurity reduction achieved by splitting on that feature at a threshold t:

$$\Delta I(m, j, t) = I(m) - \left(\frac{N_L}{N_m} I(m_L) + \frac{N_R}{N_m} I(m_R) \right), \quad (2)$$

where N_m, N_L, and N_R are the number of samples in the parent and the left and right child nodes, respectively. The optimal split at the node m is determined by evaluating all available features (or a subset of them, in the case of random forests) and all possible thresholds for each of those features to maximize ΔI. The overall importance of feature j in a single tree T is the sum of impurity reductions at all nodes m where feature j was used for splitting, weighted by the fraction of samples reaching that node.

$$GI_T(j) = \sum_{m \in Nodes(T)} (N_m/N) \Delta I(m). \quad (3)$$

For an ensemble or random forest, this is averaged over all trees.

$$GI_{MDI}(j) = \frac{1}{K} \sum_{T=1}^{K} GI_T(j). \quad (4)$$

Features yielding larger impurity reductions across many trees are considered more important, so Gini importance can be used as a feature selection technique, where features with low Gini importance scores can be removed without significantly impacting model performance. The Gini importance is computationally efficient to compute as it is a byproduct of training. This makes it a fast and simple method for assessing feature importance. However, it is calculated on the training data which means that it can overestimate the importance of features with many possible values (high cardinality), as they have more opportunities to create splits. It may also not accurately reflect the true importance of correlated features, as it might overestimate the importance of one feature while neglecting the others. This reflects the feature's role in the model's internal structure rather than directly measuring its impact on predictive performance on unseen data.

4.1.2. Permutation Importance

Permutation importance (PI), also known as mean decrease in accuracy/score (MDA), is a model-agnostic technique applicable to any fitted model after training on unseen data (validation or test set). It quantifies the importance of individual input features by measuring the decrease in model performance when a feature's values are randomly shuffled. The intuition is that if a feature is important for the model's predictions, then scrambling its values should significantly degrade performance. The importance $I(j)$ of feature j is the difference between the baseline score and the score after permutation:

$$PI(j) = S(D) - S(D_j) \quad (5)$$

where S is the baseline baseline performance score (e.g., accuracy, F1-score, R^2, mean squared error [73]) on the validation or test dataset D, and D_j is the dataset where the j-th feature column has been randomly permuted across samples. This process is often repeated multiple times with different random shuffles, and the importances are averaged for stability. A large drop $PI(j)$ indicates that the model heavily relies on feature j and that it is most important for prediction on unseen data. As this is applied on unseen data, it is considered more reliable than Gini importance and it can capture interaction effects implicitly. However, it is computationally more expensive as it requires multiple model inferences per feature and its interpretation can be complicated if input features are highly correlated, as permuting one feature may not change performance much because correlated features still contain the same information.

Feature importance measures, like Gini importance and permutation importance, are helpful in astronomical discovery and can play a role in anomaly detection. When building models to predict astronomical properties, feature importance helps to verify if the model relies on features expected to be important based on physics. For instance, if a photometric redshift model ranks specific color indices known to correlate strongly with redshift as most important, it increases confidence that the model has learned astrophysically meaningful relationships. Conversely, if a model achieves high accuracy and ranks an unexpected feature (e.g., a less commonly used color, a subtle morphological parameter) as highly important, it can prompt new scientific questions. This might suggest that this feature contains more predictive power than previously thought for that specific task or dataset, potentially leading to investigations into underlying physical reasons or revealing biases in the data. This can be a pathway to discovery by highlighting non-obvious correlations. Knowing which features are most predictive can guide astronomers in designing future surveys (e.g., prioritizing certain filters) or in refining feature extraction processes for future models. If a model performs well but relies heavily on features suspected to be related to instrumental effects or artifacts, feature importance can also flag this issue, preventing spurious scientific claims. The caveat is that feature importance primarily shows correlation and predictive power within the context of the specific model and data, not necessarily fundamental causation (Figure 1). GI and PI can be divergent; the former measures how effectively a feature is used to reduce impurity in splits in the training data, so it is reflective of the model's construction, whereas the latter directly measures a feature's contribution to the model's predictive performance on unseen data. However, this information is still a valuable input for scientific reasoning and hypothesis generation. Hoyle et al. [74] found that selecting the most important photometric features and adding them to standard inputs significantly improves the accuracy and reduces catastrophic outliers in machine learning-based redshift estimations for galaxies. Gini importance has been used to show that C-class flare percentages and maximum X-ray flux are particularly critical features for solar flare forecasting [75], and permutation feature importance applied to light curve characteristics of variable stars has revealed that the importance of specific features not only depends on the classification task but also on the distance metric used [76]. Using Gini importance, [77] found that spherical overdensities, as opposed to the ellipticity and prolateness (tidal shear features), are the most important features in predicting dark matter halo masses.

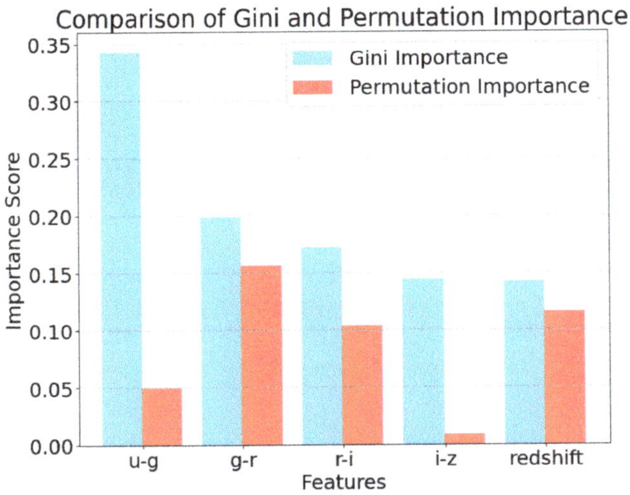

Figure 1. A visual comparison of Gini importance and permutation importance for a set of astronomical features used to classify star burst vs. star forming galaxies in the Sloan Digital Sky Survey (SDSS) DR18 data using a random forest model. Here, Gini seems to favor the "u-g" feature as the most important by a large margin, whilst the PI score suggests the contrary. Nonetheless, PI is not able to identify a single-feature dominance. This highlights how different methods can yield different interpretations.

4.2. Saliency-Based Methods

Deep learning models, particularly CNNs for image-based tasks and recurrent neural networks (RNNs) for sequential data, have become highly effective tools in astronomy for tasks like galaxy morphology classification, strong lens detection, or light curve analysis (e.g., [55,78–80]). These models, while powerful, are notoriously uninterpretable. A popular option involves generating saliency maps [81]. These methods aim to highlight which parts of the input (e.g., pixels in an image) were most influential in determining a specific output (e.g., the classification score for a particular class; see Figure 2). Techniques like integrated gradients [82] or gradient-weighted class activation mapping (Grad-CAM) [83] allow for direct comparison with known astrophysical features. In astronomy, this is especially invaluable for determining whether a model focuses on physically meaningful structures such as spiral arms, galactic bars, tidal disturbances, lensed arcs, or specific features in time series data represented as images, or are instead relying on spurious artifacts, noise patterns, or unexpected regions, potentially indicating overfitting.

In doing so, they help establish trust in the model's predictions and reveal opportunities for scientific discovery, particularly when unexpected regions emerge as influential. They also serve a critical diagnostic function; for example, when a model produces incorrect predictions, saliency maps can expose the regions responsible, assisting in model debugging and identifying edge cases. Moreover, comparisons between models with different architectures can uncover divergent decision pathways for the same task, enriching our understanding of how model design affects interpretability.

Although often compared to attention mechanisms (Section 4.4.2), saliency methods differ in that they typically operate post hoc and offer less direct insight into internal model dynamics. They show what regions in an input affect a specific output, rather than where the model focuses on. Thus, interpreting saliency maps as full causal explanations of model behavior requires caution. Nonetheless, when integrated with astrophysical domain knowledge, saliency methods contribute not only to model validation but also to the discovery of rare or subtle phenomena that might otherwise be overlooked. Jacobs et al. [85] argue the limitation of saliency maps in a scientific context as they primarily focus on identifying important spatial regions in an input image and often lack granularity or quantitative insight into model biases and weaknesses related to other physical parameters.

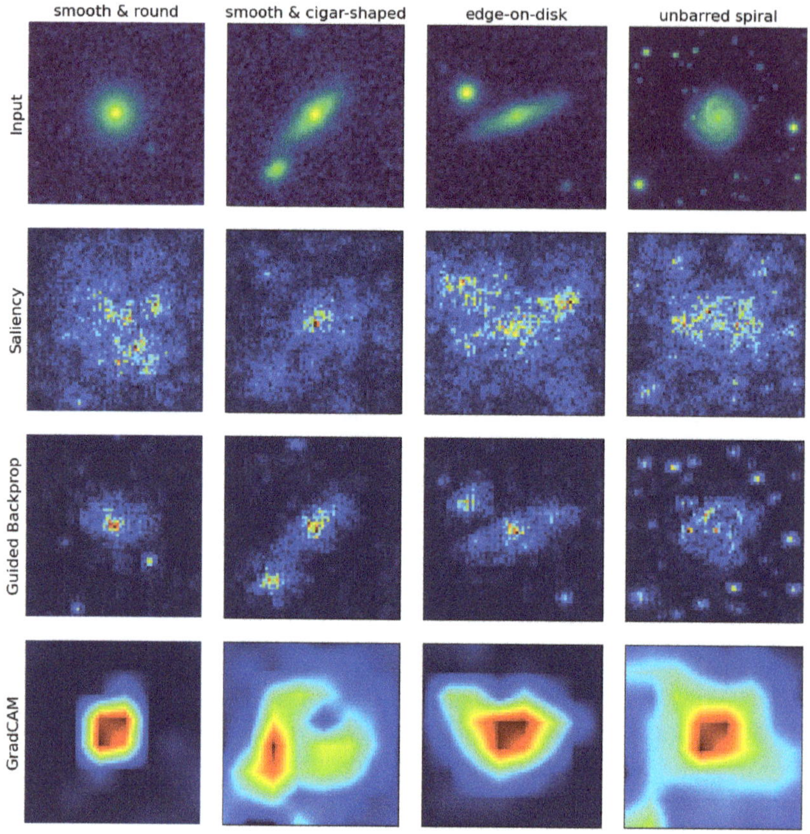

Figure 2. A comparison of saliency methods applied to a CNN trained on the Galaxy MNIST image dataset [84] to classify various morphologies of galaxies ("smooth and round", "smooth and cigar-shaped", "edge-on-disk", "unbarred spiral"). The (**top row**) shows the input images, and subsequent rows display different saliency map visualizations: vanilla saliency (**second row**), guided backpropagation (**third row**), and Grad-CAM (**bottom row**). These methods highlight the image regions most influential in the model's classification, offering insights into whether the CNN focuses on astrophysically relevant features or potential artifacts. While vanilla saliency maps provide a noisy interpretation of a pixel-level importance, GradCAM maps, derived from the last convolutional layer, are too coarse, losing the fine-grained structure, e.g., the individual surrounding galaxies. Guided backpropagation provides an intermediatry of the two.

4.2.1. Vanilla Saliency

The most basic saliency method computes the gradient of the model's output score (typically the pre-activation score for the predicted class) with respect to each input pixel x_{ij}:

$$S_{ij} = \left| \frac{\partial f(x)}{\partial x_{ij}} \right|. \tag{6}$$

where $f(x)$ is the model. The magnitude of the gradient indicates sensitivity—pixels where small changes would most affect the output score are deemed important. While simple, these maps can be noisy and suffer from gradient saturation issues where the gradient can vanish or explode.

4.2.2. Guided Backpropagation or SmoothGrad

To reduce the noise present in standard gradient saliency maps, guided backpropagation [86] modifies the gradient computation by suppressing negative gradients at ReLU activation layers during the backward pass. This adjustment focuses the explanation on features that positively contribute to the model's decision:

$$\frac{\partial f(x)}{\partial x_{ij}}|_{guided} = \text{ReLU}\left(\frac{\partial f(x)}{\partial x_{ij}}\right) \tag{7}$$

Alternatively, SmoothGrad reduces the noise by adding noise to the input image. This results in cleaner, more focused saliency maps but can lack a clear probabilistic interpretation and and may suppress important negative evidence by design.

4.2.3. Integrated Gradients

A more theoretically robust method designed to overcome saturation issues and satisfy desirable axioms like sensitivity and completeness is integrated gradients (IGs) [82]. Instead of just using the local gradient at the input x, IG integrates the gradients along a straight-line path from a chosen baseline input x' (e.g., a black image, an average image, or an image with random noise) to the actual input x. The attribution (importance score) for the i-th input feature (pixel) x_i is defined as follows:

$$IG_i(x) = (x' - x)\int_{\alpha=0}^{1} \frac{\partial f(x' + \alpha(x - x'))}{\partial x_i} d\alpha \tag{8}$$

In practice, this integral is typically approximated numerically. A key property is completeness: the sum of integrated gradients across all input features equals the difference between the model's output score at the input x and the baseline x', but while providing more reliable attributions, the choice of baseline x' can significantly influence the resulting saliency map.

4.2.4. Grad-CAM (Gradient-Weighted Class Activation Mapping)

Distinct from methods that directly attribute importance to input pixels via gradients, gradient-weighted class activation mapping (Grad-CAM) [83] is specifically tailored for CNNs. It produces coarser, heatmap-style visualizations localized using the feature maps of an intermediate or deep convolutional layer (often the final one). Grad-CAM identifies image regions contributing to the prediction of a specific class c based on the activations in that layer. If A_k is the k-th feature map of a chosen convolutional layer, then the weight is obtained by global average pooling the gradients,

$$\alpha_k^c = \frac{1}{Z}\sum_i\sum_j \frac{\partial f(x)}{\partial A_{ij}^k}. \tag{9}$$

where Z is the number of pixels in the feature map. This represents the importance of each feature map k for class c. These weights can then be used to compute a heatmap,

$$L_{Grad-CAM}^c = \text{ReLU}\left(\sum_k \alpha_k^c A^k\right) \tag{10}$$

which highlights regions important for class c at the spatial resolution of the chosen feature map. The ReLU ensures that only features positively correlated with the class are visualized. It is typically upsampled to overlay on the original image to provide spatially coarse but semantically meaningful explanations, linking output decisions to higher-level features. Grad-CAM is especially effective in identifying higher-level structures, such as galaxy morphology or strong lensing features. However, the quality of the explanation depends on the selected convolutional layer, and different choices may lead to different interpretations. Guided GradCAM [87] combines the strengths of Grad-CAM and guided backpropagation, by performing an element-wise multiplication of their outputs. The intention is to leverage the class-discriminative localization ability of Grad-CAM as a mask for the fine-grained guided backpropagation map. This should result in a higher resolution visualization. Methods like Grad-CAM have already been widely used for interpretation of CNN models in astronomy (e.g., [57,88,89]).

While saliency methods offer intuitive visual explanations, their reliability has been increasingly questioned. Adebayo et al. [90] demonstrated that many gradient-based methods, including guided backpropagation, produce visually similar maps even when the model's learned parameters are randomized, suggesting that these methods might be dominated by input processing such as edge detection rather than reflecting learned model knowledge. Similarly, Srinivas and Fleuret [91] argued that gradient-based explanations may primarily highlight regions causing maximal change rather than those representing meaningful semantic features used by the model. These findings highlight the importance of applying sanity checks and being cautious about over-interpreting saliency maps without rigorous validation, especially when such methods are used in scientific contexts.

4.3. *Model Agnostic Methods*

Model-agnostic tools like SHAP (SHapley Additive Explanations) [34] and LIME (Local Interpretable Model-Agnostic Explanations) [33] provide versatile frameworks for explaining predictions across a wide range of model types. SHAP values offer a theoretically grounded decomposition of predictions into feature contributions, while LIME fits simple surrogate models in the local neighborhood of a prediction to approximate the behavior of the full model. Both are widely used to evaluate feature relevance, detect biases, and facilitate model debugging, though they come with limitations.

4.3.1. SHAP: SHapley Additive Explanations

SHAP values (Figure 3) are based on the concept of Shapley values from cooperative game theory [92]. Given a function f representing the model's prediction, the goal is to assign a value ϕ_i to each feature x_i such that

$$f(x) = \phi_0 + \sum_{i=1}^{n} \phi_i, \tag{11}$$

where ϕ_0 is the expectation value (i.e., the baseline prediction when no features are present), and each ϕ_i represents the contribution of feature i to the prediction for input x.

Formally, the Shapley value for feature i is defined as

$$\phi_i = \sum_{S \subseteq N\{i\}} \frac{|S|!(|N| - |S| - 1)!}{|N|!} (f_x(S \cup \{i\}) - f_x(S)) \tag{12}$$

where N is the set of all the features, and $f_x(S) = E[f(X)|X_S = x_S]$ is the expected output of the model conditional on knowing the values x_S for features in subset S. This expectation is often approximated using a background dataset or other techniques. $f_x(S \cup \{i\})$ is similarly defined when feature i is also known. The term $[f_x(S \cup \{i\}) - f_x(S)]$ represents the marginal contribution of adding feature i to the subset S. The combinatorial term weights this contribution according to its position in all possible feature orderings. While SHAP is primarily designed for local interpretability (explaining individual predictions), it can also produce global explanations by aggregating local explanations across multiple instances, visualizing them through SHAP summary and dependence plots.

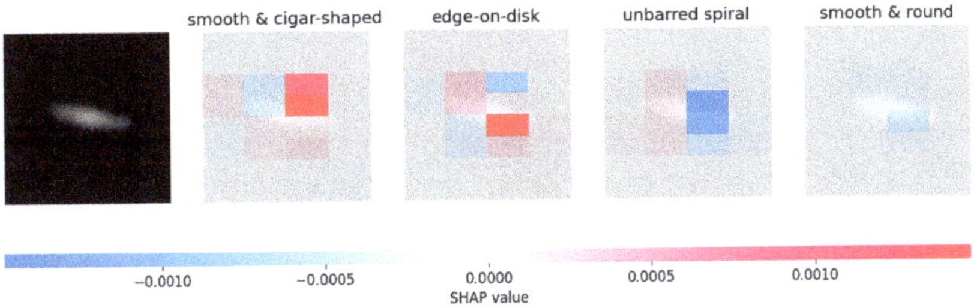

Figure 3. SHAP applied to a CNN model trained to classify galaxy morphologies. The example shows the explanation of a correctly predicted smooth-cigar-shaped galaxy. Here, superpixels with positive SHAP values (red) contribute positively towards the prediction. Blue contribute negatively. We use the partition explainer, which does not assume that features are independent from each other. The SHAP explanation suggests that the focus is on the center of the image, with the disk contributing negatively towards the spiral or round classes.

SHAP values provide feature attributions satisfying several desirable properties that contribute to their theoretical appeal. These include efficiency (the sum of feature contributions equals the difference between the prediction and the baseline), symmetry (features contributing identically receive the same attribution), the dummy property (features having no effect on the prediction receive zero attribution), and additivity (the SHAP values for combined models are the sum of the values from individual models). However, computing exact Shapley values is typically computationally intractable for models with many features. Consequently, practical SHAP implementations rely on efficient approximations or model-specific algorithms, such as the popular TreeSHAP for tree-based ensembles, to estimate these values efficiently. SHAP has been used to understand molecular abundances in star-forming regions. Heyl et al. [93] were able to quantify parameters on molecular abundances; for example, they discovered that H_2O and CO's gas phase abundances depend strongly on the metallicity, as well as reconfirming other known relationships with abundances, and Ye et al. [94] used SHAP to identify five key absorption features in spectra for carbon star identification.

4.3.2. LIME: Local Interpretable Model-Agnostic Explanations

LIME (Figure 4) takes a different approach operating on a simple, intuitive principle to approximate the behavior of a complex, potentially non-linear model f in the local neighborhood of a specific input instance x using a simpler, inherently interpretable surrogate model g (e.g., a sparse linear model or a shallow decision tree). For a linear model, the explanation g for an instance z' (a simplified, interpretable representation of z) takes the form

$$g(z') = w_g z', \tag{13}$$

where w_g are the weights (feature importances) to be learned. LIME constructs a local dataset by perturbing the original input and observing the model's response:

$$\mathcal{L}(f, g, \pi_x) = \sum_{z \in Z} \pi_x(z)(f(z) - g(z'))^2 + \Omega_g \tag{14}$$

This loss function measures how well g approximates f for the perturbed samples z weighted by their proximity π_x, which defines the neighborhood around x. To ensure that the explanation g remains interpretable, a complexity penalty Ω_g is added. For linear models, this might encourage sparsity (e.g., minimizing the number of non-zero weights or using an L1-norm penalty). For decision trees, it might penalize depth. The surrogate model g is trained to minimize $\mathcal{L}$, yielding feature weights that approximate the influence of each input feature locally. Unlike SHAP, LIME does not require exhaustive enumeration of feature subsets and is computationally more efficient, but its explanations can vary depending on the sampling process and surrogate complexity. In classifying galaxy morphologies, Goh et al. [95] showed that LIME enabled the effective identification of influential regions in the image but also showed that the model utilized some unexpected image regions beyond the galaxy object itself for classification, potentially an area of bias inherent in the images.

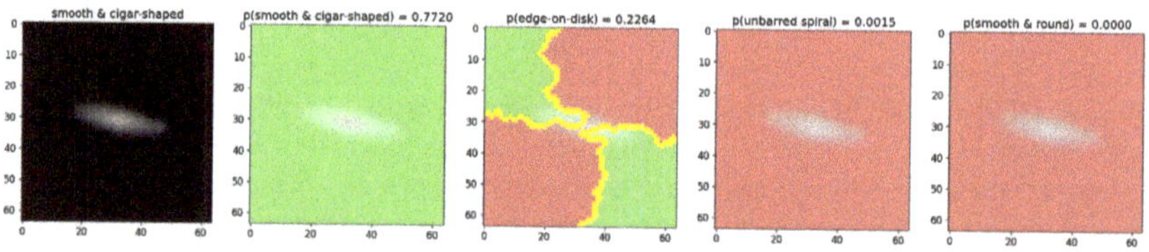

Figure 4. LIME explanation of a correctly predicted smooth-cigar-shaped galaxy by a CNN classifier model. Green corresponds to superpixels that positively contribute to the prediction, and red corresponds to superpixels that negatively contribute. This highlights that the LIME interpretation is that the classification focuses on the image as a whole rather than any particular region.

4.4. *Interpretable Models by Design*

While post hoc explanations for "black-box" models, particularly deep neural networks, are widely used ultimately, they are based on approximations that can be misleading. An alternative and growing field focuses on designing model architectures that are inherently structured for transparency, facilitating direct interpretation (e.g., linear regression, decision trees). In these models, the explanations are the model itself, offering stronger foundations for trust, transparency, and fairness that is especially critical in areas where the decision stakes are high. However, it is important to remember the trade-off between interpretability and accuracy, especially in high-dimensional, non-linear tasks. Inherently interpretable models are typically more limited (e.g., simpler models), and performance accuracy is often lower on complex tasks. We now explore some of these approaches in more detail.

4.4.1. Rule-Based Methods

Feature importance techniques (like Gini or permutation importance) tell us which features the model found most influential in general. Rule-based approaches represent a class of inherently interpretable models that aim for complete transparency in their decision-making process through their structure. The most common example is the decision tree [96], where predictions are made by following a specific path of IF–THEN conditions from a root node to a leaf node, based on thresholds applied to input feature values (Figure 5).

This explicit logic chain makes it straightforward, in principle, to understand exactly how a specific prediction was derived for any given input instance. Other methods, like rule lists or algorithms such as RuleFit [97], explicitly generate sets of rules often combined with linear models.

The primary strength of these methods lies in their high interpretability and transparency. The decision logic is human-readable, making them intuitive and particularly useful in classification tasks requiring well-defined decision boundaries or where explaining the reasoning to stakeholders is paramount. They can also naturally capture interaction effects between features within a single rule path. But, despite their transparency, rule-based methods often have limited flexibility and expressive power, particularly when dealing with the complex, high-dimensional, and often noisy data typical of modern astronomical surveys where intricate non-linear relationships may exist. Very deep decision trees, while technically transparent, can become visually complex and difficult for humans to fully grasp. Furthermore, tree structures can be unstable. Small variations in the training data can lead to significantly different trees and rules. Consequently, they may achieve lower predictive accuracy compared to ensemble methods or deep learning models on many complex astronomical tasks.

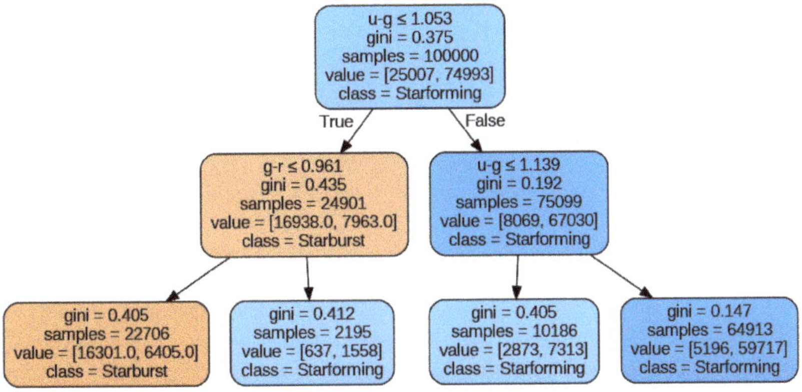

Figure 5. The structure of a decision tree classifier applied to classify and star-forming galaxies in the Sloan Digital Sky Survey (SDSS) DR18. Each internal node displays the feature threshold (e.g., color "u-g $\leq$ 1.053" at the root), the Gini impurity, the number of samples reaching that node, the distribution of samples across classes (value), and the majority class at that point. For instance, the root node starts with 100,000 samples and splits them based on the "u-g" color. If the condition is "True", samples proceed to the left child (24,901 samples, majority class "Non-Star-forming") and if "False", they go to the right child (75,099 samples, majority class "Star-forming"). The tree continues to split samples, aiming to create purer leaf nodes that represent the final classifications. By observing feature importance and visualizing the decision tree, we gain insights into how the model classifies galaxies based on the chosen features and their ranges. The first split on "u-g" implies that this color magnitude plays a significant role in the classification.

While perhaps less frequently deployed for achieving state-of-the-art performance on complex raw data analysis today, rule-based logic remains conceptually valuable. Simple decision trees have been used for basic classification tasks (e.g., star/galaxy separation based on magnitude and morphology metrics (e.g., [98])) or formed components of early transient alert systems [99]. More complex rule-based classification algorithms like random forests can effectively learn the complex relationship between the initial density field conditions of dark matter particles and their final state (whether they end up in massive halos), allowing for accurate prediction of simulation outcomes without running the full simulation [100].

4.4.2. Attention Mechanisms

Often compared to saliency are attention mechanisms. Initially developed to improve performance in tasks like machine translation [101], they have been increasingly integrated into neural architectures [35]. They offer a natural interface for interpretability beyond simply analyzing feature rankings or explicit rules by highlighting which parts of the input (e.g., image regions, sequence elements, specific features) the model focuses on when generating an output.

The core idea behind attention is to allow a model to dynamically weigh the importance of different parts of its input data, rather than relying solely on fixed-size receptive fields (as is standard in CNNs) or compressing all prior information into a single hidden state vector (as in basic RNNs). When processing an element (e.g., predicting the next word in a sentence, classifying an image), the attention mechanism computes a set of weights called attention scores over the input elements, indicating how much attention or importance should be given to each one.

A dominant form of attention, particularly central to the influential Transformer architecture [35], is scaled dot-product attention, also known as self-attention (Figure 6). Mathematically, the self-attention mechanism computes attention scores as follows:

$$\alpha = softmax\left(\frac{Q \cdot K}{\sqrt{d}}\right) \quad (15)$$

With the final weighted output representation:

$$Attention(Q, K, V) = \alpha V \quad (16)$$

where the queries (Q), keys (K), and values (V) are vectors derived from the input embeddings. Query (Q) represents the current element or context asking for information ("What am I looking for?"). Key (K) represents identifiers or labels for the input elements ("What information does this element have?"), and value (V) represents the actual content or features of the input elements associated with the keys. The dimensionality of the key vectors d is used for scaling, and the resulting attention matrix α is often treated as a measure of interpretability. It can be visualized to inform use about which input tokens or spatial regions in an image the model focused on.

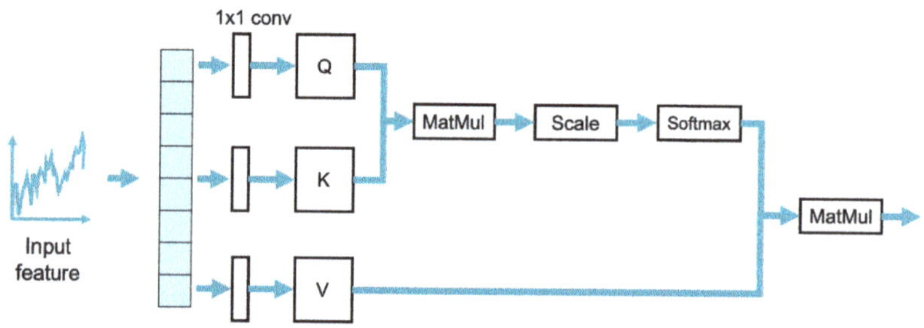

Figure 6. Scaled dot-product attention mechanism.

Self-attention computes the relevance of different parts of the input sequence to each other. But a single attention layer might only capture one type of relationship (e.g., similarity in spatial location or color in an image, or proximity in time in a light curve). To address this, multi-head attention runs several attention mechanisms in parallel ("heads"), each with its own learnable parameters, allowing the model to jointly attend to information from different representation subspaces at different positions. The outputs from all heads are concatenated and projected.

Attention mechanisms are inherently more interpretable than traditional deep learning layers, because they produce explicit, learnable weights that can be visualized and analyzed (Figure 7). These attention scores can be mapped back to domain-specific structures: pixels in an image, time points in a signal, or wavelengths in a spectrum—offering intuitive insights for scientists. Like feature importance, attention is not an anomaly detector, per se, but when a model flags a galaxy or transient as unusual, the attention map can reveal which regions of the image or spectral range triggered the anomaly, and whether the model is focusing on meaningful astrophysical structure or noise/artifacts. This grounds the anomaly score in interpretable terms. Attention maps can also help filter false positives by revealing when the model's focus is on irrelevant regions (e.g., image borders, artifacts). However, a significant body of research argues that attention weights are not a direct or reliable explanation of model predictions [102]. Reasons include that high attention does not guarantee causality, the complex non-linear transformations after the attention layer can influence the final prediction, and the attention weights can sometimes be manipulated without significantly changing the model's output, whilst equally significant are challenges against those claims [103]. In any case, while attention maps offer valuable insight into the model's internal processing and where it looks, they should be interpreted with caution. They are a potentially useful diagnostic tool but not necessarily a faithful explanation of why a prediction was made. Ref. [104] used attention-gating not only to improve their classification of radio galaxies but also to help choose models that align better with how astronomers classify radio sources by eye.

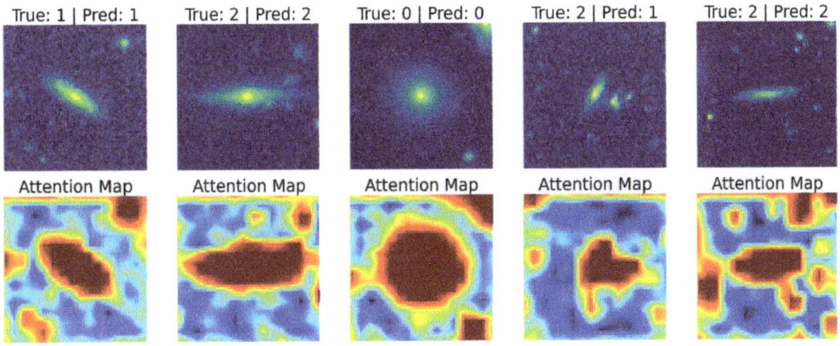

Figure 7. Attention maps of a CNN with attention model trained on the Galaxy MNIST image dataset. The (**top row**) shows the input galaxy images with the true and predicted classes. The (**bottom row**) shows the corresponding attention maps with red indicating high attention. While the attention maps highlight visible galaxy structures, they can also focus on regions less obvious to humans, such as fainter outskirts or surrounding galaxies. For instance, neighboring galaxies may be highlighted if they are involved in mergers (affecting morphology) or if they are indicative of environmental density, as galaxy morphology often correlates with environment (e.g., ellipticals in clusters, spirals in the field). These maps indicate computationally influential regions that may not perfectly align with human-perceived salient astrophysical features and are worth further investigation.

4.4.3. Symbolic Regression

Symbolic regression is an interpretable machine learning technique that seeks to discover analytical expressions that best model a given dataset, without assuming a predefined functional form (like linear regression or neural networks). Unlike traditional regression methods that fit coefficients to a fixed equation, symbolic regression algorithms explore a combinatorial space of mathematical operators (e.g., +, $-$, $\times$, $\div$, exp, log) and input features to construct candidate models [105]. The objective is to minimize a loss function $\mathcal{L}(y, \hat{y})$ while simultaneously optimizing for simplicity and parsimony, often guided by multi-objective optimization:

$$\min_{f \in F}(\mathcal{L}(y, f(x)) + \lambda Complexity(f)), \tag{17}$$

where F denotes the space of symbolic expressions, and $Complexity(f)$ is a regularization term penalizing complexity.

The primary appeal of SR lies in its potential for high interpretability and transparency. The direct output is a human-readable mathematical formula, representing perhaps the ultimate form of a "white-box" model. If the discovered equation is compact, accurate, and physically plausible, it can provide profound scientific insight, potentially revealing previously unknown empirical relationships or even approximations to fundamental physical laws directly from data [106,107]. This contrasts sharply with black-box models where the learned relationships are opaque. This approach is particularly compelling in physics-informed contexts, where the recovery of closed-form expressions aligns with scientific intuition and the formulation of empirical or theoretical laws. SR can generate simple, interpretable analytical approximations for complex physical simulations or theoretical functions where a readily understandable formula is desired [108] and in astronomy, symbolic regression has been used to rediscover known physical laws from simulated data and holds promise for uncovering novel empirical scaling relations between galaxy properties, parameters describing stellar evolution, or equations governing orbital dynamics directly from observational or simulation data [109–111].

Despite its interpretability, SR faces several limitations. The search space is typically vast and non-convex, often requiring evolutionary algorithms (e.g., genetic programming) or more recent neural-guided approaches (e.g., deep symbolic regression) to navigate efficiently. As a result, symbolic regression can be computationally expensive, especially in high-dimensional settings where many candidate expressions must be evaluated. Additionally, symbolic models may be sensitive to noise, and the lack of strong priors can lead to overfitting or implausible expressions in data-poor regimes.

Nevertheless, SR offers a rare combination of accuracy and explicitness that makes it a valuable tool for hypothesis generation and scientific insight in astronomy. Its outputs are interpretable, closed-form equations that can be scrutinized, validated, or rejected in the light of physical principles, fostering a deeper understanding of the patterns uncovered by machine learning. Wadekar et al. [112] trained a neural network to predict neutral hydrogen content from dark matter fields. The saliency maps revealed that the neural network considered the halo's environment, not just the halo itself, when making predictions, motivating the closer exploration of assembly bias. Using SR, they were able to parameterize a new and physically interpretable model of assembly bias.

4.4.4. Learning Interpretable Latent Representations

In some machine learning models, particularly those involving dimensionality reduction like autoencoders, the input data are transformed into a latent space. This space is a lower-dimensional, abstract representation where the original data are encoded, often in a compressed form. While this encoding is structured, it is not automatically understandable by humans unless the model is specifically designed to be so. Interpretability in latent spaces means understanding what each latent variable encodes, with, ideally, each dimension corresponding to a distinct, meaningful factor of variation (e.g., one variable controls rotation, another controls scale). This can be encouraged through the following:

- Disentanglement techniques, which aim to separate independent factors of variation in the latent representation by adding additional constraints to the loss function to encourage each latent dimension to capture an independent aspect of the data. Models like β-variational autoencoders (β-VAEs) or other disentangled VAEs perform this. The β-VAE loss is defined as

$$L = L_{reconstruction} + \beta D_{KL}(q_\phi(z|x)||p(z)) \quad (18)$$

where $L_{reconstruction}$ is the original loss, $D_{KL}(q_\phi(z|x)||p(z))$ is the Kullback–Leibler (KL) divergence that measures how much the learned latent distribution $q_\phi(z|x)$ deviates from a chosen prior distribution $p(z)$ (typically a standard Gaussian), and the hyperparameter β encourages the model to learn a latent distribution that is closer to the simple prior and, thus, promotes disentanglement.

- Conditional generation models like conditional variational autoencoders (CVAEs) condition the model on known variables like class labels. By observing how the latent space changes when conditioned on different known variables, its possible to infer how certain learned features in the latent space relate to these explicit conditions.
- Latent traversals involve systematically changing one latent variable at a time (while keeping others fixed) and observing the generated outputs. This technique can reveal what each dimension represents and whether it aligns with human-understandable concepts.
- Post hoc analysis of latent space structure: After training, dimensionality reduction techniques like principal component analysis (PCA) and clustering on the latent space can be an effective way to further explore the latent space to gain insights to what the model has learned.

An application of these principles is demonstrated by [113] who trained a variational encoder using a β-VAE loss to predict the halo mass function (HMF) by compressing it into a three-dimensional latent space. Post training, they use mutual information analysis to understand the cosmological dependence of each latent variable. Notably, one latent variable captured non-universal HMF behavior, linking it to the Universe's recent growth history after matter–dark-energy equality, suggesting that subtle differences in the Universe's expansion history after dark energy becomes dominant, leading to the deviations from universality.

However, learning latent variables that align with human-understandable concepts remains challenging [114]. Even disentangled latent variables might not consistently map semantic features, and for models like VAEs, the latent space is probabilistic (variables are samples from distributions). This inherent randomness can add a layer of complexity when trying to interpret the precise meaning of the encoding for individual data points.

4.4.5. Physics-Informed Neural Networks (PINNs)

Whilst not intrinsically interpretable in the same way that simpler models like linear regression or decision trees are, physics-informed neural networks (PINNs) possess characteristics that make them more interpretable than standard "black-box" neural networks. These networks incorporate knowledge of physical laws, often as partial differential equations (PDEs), into the loss function used to train the neural network [115]. This architectural bias ensures that the model predictions remain consistent with established physical principles. We can interpret their outputs in the context of these laws, and we can trust that the model's behavior is, to some extent, governed by these principles, enhancing their transparency and trustworthiness. By encoding domain knowledge directly into the model structure, PINNs help bridge the gap between black-box neural networks and physically interpretable modeling; however, despite their promise, challenges remain in training efficiency, handling stiff PDEs, and scaling to high dimensions [116].

4.5. *Prototypes and Exemplars*

While many interpretability methods are benchmarked on supervised tasks due to the presence of a ground truth, understanding the outputs of unsupervised methods like clustering is also needed. Here, interpretability often shifts to understanding the defining characteristics of the discovered groups or structures and assessing their astrophysical relevance, rather than explaining a singular prediction. This can be challenging as there is no ground truth and, hence, no explanation target, and quantitative evaluation of interpretation correctness is less straightforward. An intuitive approach is then reasoning by analogy, explaining predictions by comparing new inputs to known examples from the training set. These prototype- and exemplar-based methods offer explanations by referencing representative data points from the training set, grounding a model's decision to specific instances rather than abstract features or internal model mechanics [117]. They operate on the principle "This input is class Y because it resembles these known examples of Y", requiring a meaningful similarity function defined either in the original feature space or a learned embedding space.

Two main variants exist. Exemplar-based methods, like k-nearest neighbors, use actual training data points, making predictions directly traceable to real observations. Prototype-based methods instead use learned or synthetic representatives (e.g., cluster centroids or learned latent prototypes) that act as archetypes for a class, summarizing groups of examples. Unsupervised learning techniques [70] are frequently used to discover or define potential prototypes or to create the feature space in which exemplars are identified. The appeal lies in their alignment with human reasoning, mirroring how astronomers interpret visual data (e.g., classifying galaxies by comparison, as in Galaxy Zoo) or time series data (e.g., comparing supernova light curves). This approach is also useful for model debugging, allowing inspection of influential examples to diagnose issues like overfitting or dataset biases.

However, limitations exist. Effectiveness hinges on the quality and interpretability of the chosen similarity metric, as proximity in high-dimensional or latent spaces may not reflect true semantic similarity. These methods typically provide only local explanations, can be computationally demanding for large datasets, and often lack feature level granularity, explaining that an input is similar but not which features drive the similarity.

Despite these challenges, prototype- and exemplar-based explanations remain valuable, particularly when intuitive, traceable interpretations grounded in real data are needed. In astronomy, where expert visual inspection is common, they help bridge the gap between complex model outputs and scientific understanding.

4.6. AI Reasoning Models

One area that is seeing significant benefits of xAI is, surprisingly, large language models (LLMs). Traditional LLMs excel at generating text and answering questions, but they often struggle with tasks requiring logical deduction or multi-step problem solving. Their adoption in scientific research is, therefore, often hindered by limitations such as a tendency for generating plausible but incorrect results and a lack of true understanding of underlying principles. The lack of transparency in how these models arrive at their conclusions makes it difficult to validate their results and build trust in their predictions, especially when applied in astronomy where verifiable logic is paramount. As LLMs become integrated into scientific workflows, from data analysis to hypothesis generation, we need to ensure that we understand how they arrive at their outputs. Methods like chain-of-thought (CoT) and question–analysis prompting guide models to reason step-by-step, improving performance on complex tasks and offering a degree of process transparency [118,119]. These techniques allow users to follow and evaluate the logic behind model outputs, helping to catch errors and build trust.

However, this surface-level reasoning can mask deeper issues. CoT outputs may appear logical but do not necessarily reflect the model's true internal processes [120–122]. Caution is warranted, as CoT's seemingly logical steps may not be faithful [123], potentially representing post hoc rationalizations, flawed logic, or pattern mimicry instead of true reasoning. Faithfulness remains difficult to measure, existing metrics are debated [124], and CoT reasoning is vulnerable to adversarial attacks [125]. This highlights a broader challenge: enhancing interpretability in AI-driven science not just through better outputs, but through better understanding of model behavior. The emergence of agentic systems like the AI Scientist-v2 [126] and AI Cosmologist [127] are moving us towards fully automated research pipelines, from hypothesis generation to code implementation and analysis, further emphasizing the rapid developments in AI and the need for transparent and trustworthy reasoning for a future in astronomy that uses AI responsibly.

5. Navigating the Future and Concluding Remarks

Artificial intelligence is rapidly advancing astronomical discovery; yet, moving forward, the successful integration of it into the core of astronomical research requires more than just predictive accuracy. The ultimate scientific value of these powerful tools hinges critically on our ability to understand, validate, and trust their outputs through interpretable and explainable methodologies.

Despite progress, the practical application of trustworthy and interpretable AI in astronomy faces substantial open challenges and limitations. A primary concern is scalability; many sophisticated xAI techniques (such as model-agnostic methods like LIME and SHAP that require multiple model evaluations, or symbolic regression with its vast search space for expressions) incur significant computational costs, making their application to the petabyte- and exabyte-scale datasets from facilities like LSST or SKA, or to increasingly complex deep learning models, a major bottleneck [128], hindering their effective use in real-time transient astronomy. Fundamental questions also persist about the faithfulness of explanation and the reliability of the training data, including whether data uncertainties are adequately considered. For instance, while visually appealing, vanilla saliency maps can be noisy and sensitive to model architecture. GradCAM can be too coarse, leading to missed fine-grained features. Likewise, attention mechanisms, despite being inherent to the model, can be misleading, potentially highlighting correlations rather than causal links. Furthermore, some saliency methods have been shown to be insensitive to model parameter and label randomization, questioning their dependency on learned parameters [37,90,129]. This issue extends to other methods: LIME explanations are highly sensitive to perturbation strategy and kernel width, and its linear approximations may capture non-linear behaviors; permutation-based feature importance can be misleading for highly correlated features; and achieving true, meaningful disentanglement in latent spaces of interpretable-by-design models like β-VAEs remains notoriously difficult. Compounding this is the lack of standardized benchmarks and evaluation metrics specifically tailored for assessing explanation quality within the astronomical context [130,131]. Even when explanations are generated, their potential complexity can make them difficult for domain scientists to interpret effectively, hindering actionable insights. Given these varied limitations, it is clear that no single xAI method offers a complete understanding. Relying on just one technique risks painting a biased picture. We, therefore, encourage the reader to employ and compare a range of interpretability methods and look for consensus. If different methods highlight similar features or patterns, confidence in the interpretation increases. It is also imperative to test explanations, for example, by observing how the model's prediction changes when that feature is perturbed or removed. Furthermore, as different training runs can produce equally valid but non-unique networks, the robustness of explanations to such variations must be considered. Addressing this full spectrum of challenges necessitates developing more efficient, demonstrably faithful xAI methods, alongside tools and interfaces that lower the barrier for user expertise.

Beyond these technical hurdles lie important ethical considerations. While large public astronomical datasets generally pose minimal individual privacy risks compared to other fields [132], responsible data governance, including equitable access to data and tools, and transparency in methodology remain crucial. A significant ethical challenge is mitigating bias; AI models trained on historical astronomical data can inadvertently learn and perpetuate observational selection effects or existing societal biases present in data collection or labeling, potentially leading to skewed scientific conclusions. Interpretable AI methods are vital tools for detecting and potentially mitigating these biases, ensuring fairness and robustness in scientific findings. Responsible use also demands transparency from researchers regarding their use of AI and xAI methods, including acknowledging limitations. Over-reliance on complex models without adequate interpretation, or using explanations to simply reinforce preexisting beliefs, could undermine the scientific process itself. Addressing this full spectrum of technical, practical, and ethical challenges through continued research and interdisciplinary dialogue is essential for the responsible and effective advancement of AI-driven discovery in astronomy.

Despite the challenges, the future of interpretable machine learning in astronomy discovery looks promising. Significant advancements are being pursued in deep learning interpretability, moving beyond surface-level explanations towards mechanistic interpretability aiming to reverse engineer the specific algorithms learned within complex neural networks [133]. There is also exciting potential in leveraging ML for causal inference, enabling a shift from identifying correlations to understanding cause-and-effect relationships directly from astronomical data [134]. Furthermore, the synergy between physics-informed machine learning [135,136], which embeds physical laws into models, and xAI techniques promises models that are not only potentially more robust but also interpretable by design. Future directions also include developing hybrid AI systems (e.g., neural-symbolic) and domain-specific interpretability methods tailored to the unique characteristics of astronomical data (images, spectra, time series, simulations), alongside more robust techniques for evaluating the faithfulness and utility of explanations [137]. These trends collectively point towards a future where AI serves as a more transparent, reliable, and insightful partner in astronomical discovery.

Advancing trustworthy AI in astronomy fundamentally requires closer interdisciplinary collaboration and the establishment of shared best practices. Fostering joint workshops, dedicated funding initiatives supporting cross-domain teams, and developing common software libraries tailored for astronomical data can bridge the gap between AI researchers and domain scientists. Simultaneously, the community should work towards standardizing IML workflows: developing guidelines for the application and reporting of explainability methods in publications, creating astronomy-specific benchmarks to objectively evaluate different techniques, and promoting open-source sharing of models and interpretation code. Such concerted efforts will accelerate the development and responsible adoption of robust, reliable, and truly insightful interpretable machine learning tools for astronomical discovery.

This review surveyed the profound impact of artificial intelligence on astronomical discovery, demonstrating its necessity in the face of exponentially growing data. The central takeaway, however, is that predictive power alone is insufficient for scientific progress; it must be built upon a foundation of trust, which is cultivated through interpretability and explainability. Understanding how and why AI models arrive at their conclusions allows astronomers to validate findings, debug complex systems, guard against bias, and discover new scientific insights directly from the learned representations. We collated some key interpretation tools, but addressing the current limitations in xAI and embracing emerging trends towards more transparent, causal, and physically grounded models is a still an ongoing endeavor for the future of the field. Ultimately, through the adoption of best practices by prioritizing interpretability alongside predictive performance, trustworthy AI will become an indispensable and reliable partner for astronomical discovery.

Funding: This research received no external funding.

Data Availability Statement: The code and data used in this paper are publicly available at https://github.com/MaggieLieu/interpretable_ML_examples.git, accessed on 3 June 2025.

Acknowledgments: M.L. would like to thank the anonymous reviews for their insights into improving this manuscript. She acknowledges a research fellowship in machine learning and cosmology from the University of Nottingham.

Conflicts of Interest: The author declares no conflicts of interest.

References

1. Hubble, E.P. A spiral nebula as a stellar system, Messier 31. *Astrophys. J.* **1929**, *69*, 103–158. [CrossRef]
2. Benn, C.R.; Sanchez, S.F. Scientific impact of large telescopes. *Publ. Astron. Soc. Pac.* **2001**, *113*, 385. [CrossRef]
3. O'Riordan, C.; Oldham, L.; Nersesian, A.; Li, T.; Collett, T.; Sluse, D.; Altieri, B.; Clément, B.; Vasan, K.; Rhoades, S.; et al. Euclid: A complete Einstein ring in NGC 6505. *Astron. Astrophys.* **2025**, *694*, A145. [CrossRef]
4. Martinez-Delgado, D.; Stein, M.; Pawlowski, M.S.; Makarov, D.; Makarova, L.; Donatiello, G.; Lang, D. Tracing satellite planes in the Sculptor group: II. Discovery of five faint dwarf galaxies in the DESI Legacy Survey. *arXiv* **2024**, arXiv:2405.03769.
5. MacDonald, E.A.; Donovan, E.; Nishimura, Y.; Case, N.A.; Gillies, D.M.; Gallardo-Lacourt, B.; Archer, W.E.; Spanswick, E.L.; Bourassa, N.; Connors, M.; et al. New science in plain sight: Citizen scientists lead to the discovery of optical structure in the upper atmosphere. *Sci. Adv.* **2018**, *4*, eaaq0030. [CrossRef]
6. Lintott, C.J.; Schawinski, K.; Slosar, A.; Land, K.; Bamford, S.; Thomas, D.; Raddick, M.J.; Nichol, R.C.; Szalay, A.; Andreescu, D.; et al. Galaxy Zoo: Morphologies derived from visual inspection of galaxies from the Sloan Digital Sky Survey. *Mon. Not. R. Astron. Soc.* **2008**, *389*, 1179–1189. [CrossRef]
7. York, D.G.; Adelman, J.; Anderson Jr, J.E.; Anderson, S.F.; Annis, J.; Bahcall, N.A.; Bakken, J.; Barkhouser, R.; Bastian, S.; Berman, E.; et al. The sloan digital sky survey: Technical summary. *Astron. J.* **2000**, *120*, 1579. [CrossRef]
8. Kaiser, N.; Aussel, H.; Burke, B.E.; Boesgaard, H.; Chambers, K.; Chun, M.R.; Heasley, J.N.; Hodapp, K.W.; Hunt, B.; Jedicke, R.; et al. Pan-STARRS: A large synoptic survey telescope array. In *Proceedings of the Survey and Other Telescope Technologies and Discoveries*; SPIE: Bellingham, WA, USA, 2002; Volume 4836, pp. 154–164.
9. Ivezić, Ž.; Kahn, S.M.; Tyson, J.A.; Abel, B.; Acosta, E.; Allsman, R.; Alonso, D.; AlSayyad, Y.; Anderson, S.F.; Andrew, J.; et al. LSST: From science drivers to reference design and anticipated data products. *Astrophys. J.* **2019**, *873*, 111. [CrossRef]
10. Dewdney, P.E.; Hall, P.J.; Schilizzi, R.T.; Lazio, T.J.L. The square kilometre array. *Proc. IEEE* **2009**, *97*, 1482–1496. [CrossRef]
11. Laureijs, R.; Amiaux, J.; Arduini, S.; Augueres, J.L.; Brinchmann, J.; Cole, R.; Cropper, M.; Dabin, C.; Duvet, L.; Ealet, A.; et al. Euclid definition study report. *arXiv* **2011**, arXiv:1110.3193.
12. Gardner, J.P.; Mather, J.C.; Clampin, M.; Doyon, R.; Greenhouse, M.A.; Hammel, H.B.; Hutchings, J.B.; Jakobsen, P.; Lilly, S.J.; Long, K.S.; et al. The james webb space telescope. *Space Sci. Rev.* **2006**, *123*, 485–606. [CrossRef]
13. Spergel, D.; Gehrels, N.; Baltay, C.; Bennett, D.; Breckinridge, J.; Donahue, M.; Dressler, A.; Gaudi, B.; Greene, T.; Guyon, O.; et al. Wide-field infrared survey telescope-astrophysics focused telescope assets WFIRST-AFTA 2015 report. *arXiv* **2015**, arXiv:1503.03757.
14. Vogelsberger, M.; Marinacci, F.; Torrey, P.; Puchwein, E. Cosmological simulations of galaxy formation. *Nat. Rev. Phys.* **2020**, *2*, 42–66. [CrossRef]
15. Skoda, P.; Adam, F. *Knowledge Discovery in Big Data from Astronomy and Earth Observation: AstroGeoInformatics*; Elsevier: Amsterdam, The Netherlands, 2020.
16. Collaboration, E.; Böhringer, H.; Chon, G.; Cucciati, O.; Dannerbauer, H.; Bolzonella, M.; De Lucia, G.; Cappi, A.; Moscardini, L.; Giocoli, C.; et al. Euclid preparation. *Astron. Astrophys.* **2025**, *693*, A58. [CrossRef]
17. Narayan, G.; Zaidi, T.; Soraisam, M.D.; Wang, Z.; Lochner, M.; Matheson, T.; Saha, A.; Yang, S.; Zhao, Z.; Kececioglu, J.; et al. Machine-learning-based Brokers for Real-time Classification of the LSST Alert Stream. *Astrophys. J. Suppl. Ser.* **2018**, *236*, 9. [CrossRef]
18. Lieu, M.; Cheng, T.Y. Machine Learning methods in Astronomy. *Astron. Comput.* **2024**, *47*, 100830. [CrossRef]
19. Mehta, P.; Bukov, M.; Wang, C.H.; Day, A.G.; Richardson, C.; Fisher, C.K.; Schwab, D.J. A high-bias, low-variance introduction to machine learning for physicists. *Phys. Rep.* **2019**, *810*, 1–124. [CrossRef]

20. Graham, M.J.; Kulkarni, S.; Bellm, E.C.; Adams, S.M.; Barbarino, C.; Blagorodnova, N.; Bodewits, D.; Bolin, B.; Brady, P.R.; Cenko, S.B.; et al. The zwicky transient facility: Science objectives. *Publ. Astron. Soc. Pac.* **2019**, *131*, 078001. [CrossRef]
21. Masci, F.J.; Laher, R.R.; Rusholme, B.; Shupe, D.L.; Groom, S.; Surace, J.; Jackson, E.; Monkewitz, S.; Beck, R.; Flynn, D.; et al. The zwicky transient facility: Data processing, products, and archive. *Publ. Astron. Soc. Pac.* **2018**, *131*, 018003. [CrossRef]
22. Mahabal, A.; Rebbapragada, U.; Walters, R.; Masci, F.J.; Blagorodnova, N.; van Roestel, J.; Ye, Q.Z.; Biswas, R.; Burdge, K.; Chang, C.K.; et al. Machine learning for the zwicky transient facility. *Publ. Astron. Soc. Pac.* **2019**, *131*, 038002. [CrossRef]
23. Metzger, B.D.; Berger, E. What is the most promising electromagnetic counterpart of a neutron star binary merger? *Astrophys. J.* **2012**, *746*, 48. [CrossRef]
24. Jia, P.; Jia, Q.; Jiang, T.; Yang, Z. A simulation framework for telescope array and its application in distributed reinforcement learning-based scheduling of telescope arrays. *Astron. Comput.* **2023**, *44*, 100732. [CrossRef]
25. Adadi, A.; Berrada, M. Peeking inside the black-box: A survey on explainable artificial intelligence (XAI). *IEEE Access* **2018**, *6*, 52138–52160. [CrossRef]
26. Kamath, U.; Liu, J. Explainable Artificial Intelligence: An Introduction to Interpretable Machine Learning; Springer Nature: Berlin/Heidelberg, Germany, 2021.
27. Ćiprijanović, A.; Kafkes, D.; Snyder, G.; Sánchez, F.J.; Perdue, G.N.; Pedro, K.; Nord, B.; Madireddy, S.; Wild, S.M. DeepAdversaries: Examining the robustness of deep learning models for galaxy morphology classification. *Mach. Learn. Sci. Technol.* **2022**, *3*, 035007. [CrossRef]
28. Walmsley, J. Artificial intelligence and the value of transparency. *AI Soc.* **2021**, *36*, 585–595. [CrossRef]
29. Montavon, G.; Samek, W.; Müller, K.R. Methods for interpreting and understanding deep neural networks. *Digit. Signal Process.* **2018**, *73*, 1–15. [CrossRef]
30. Zhang, Y.; Tiňo, P.; Leonardis, A.; Tang, K. A survey on neural network interpretability. *IEEE Trans. Emerg. Top. Comput. Intell.* **2021**, *5*, 726–742. [CrossRef]
31. Gunning, D.; Stefik, M.; Choi, J.; Miller, T.; Stumpf, S.; Yang, G.Z. XAI—Explainable artificial intelligence. *Sci. Robot.* **2019**, *4*, eaay7120. [CrossRef]
32. Arrieta, A.B.; Díaz-Rodríguez, N.; Del Ser, J.; Bennetot, A.; Tabik, S.; Barbado, A.; García, S.; Gil-López, S.; Molina, D.; Benjamins, R.; et al. Explainable Artificial Intelligence (XAI): Concepts, taxonomies, opportunities and challenges toward responsible AI. *Inf. Fusion* **2020**, *58*, 82–115. [CrossRef]
33. Lundberg, S.M.; Lee, S.I. A unified approach to interpreting model predictions. In Proceedings of the Advances in Neural Information Processing Systems 30 (NIPS 2017), Long Beach, CA, USA, 4–9 December 2017.
34. Ribeiro, M.T.; Singh, S.; Guestrin, C. "Why should i trust you?" Explaining the predictions of any classifier. In Proceedings of the 22nd ACM SIGKDD International Conference on Knowledge Discovery and Data Mining, San Francisco, CA, USA, 13–17 August 2016; pp. 1135–1144.
35. Vaswani, A.; Shazeer, N.; Parmar, N.; Uszkoreit, J.; Jones, L.; Gomez, A.N.; Kaiser, Ł.; Polosukhin, I. Attention is all you need. In Proceedings of the Advances in Neural Information Processing Systems 30 (NIPS 2017), Long Beach, CA, USA, 4–9 December 2017.
36. Huppenkothen, D.; Ntampaka, M.; Ho, M.; Fouesneau, M.; Nord, B.; Peek, J.E.; Walmsley, M.; Wu, J.F.; Avestruz, C.; Buck, T.; et al. Constructing impactful machine learning research for astronomy: Best practices for researchers and reviewers. *arXiv* **2023**, arXiv:2310.12528.
37. Rudin, C. Stop explaining black box machine learning models for high stakes decisions and use interpretable models instead. *Nat. Mach. Intell.* **2019**, *1*, 206–215. [CrossRef] [PubMed]
38. Rudin, C.; Chen, C.; Chen, Z.; Huang, H.; Semenova, L.; Zhong, C. Interpretable machine learning: Fundamental principles and 10 grand challenges. *Stat. Surv.* **2022**, *16*, 1–85. [CrossRef]
39. Ntampaka, M.; Vikhlinin, A. The importance of being interpretable: Toward an understandable machine learning encoder for galaxy cluster cosmology. *Astrophys. J.* **2022**, *926*, 45. [CrossRef]
40. Schneider, P.; Ehlers, J.; Falco, E.E.; Schneider, P.; Ehlers, J.; Falco, E.E. *Gravitational Lenses as Astrophysical Tools*; Springer: Berlin/Heidelberg, Germany, 1992.
41. Treu, T. Strong lensing by galaxies. *Annu. Rev. Astron. Astrophys.* **2010**, *48*, 87–125. [CrossRef]
42. Collett, T.E.; Oldham, L.J.; Smith, R.J.; Auger, M.W.; Westfall, K.B.; Bacon, D.; Nichol, R.C.; Masters, K.L.; Koyama, K.; van den Bosch, R. A precise extragalactic test of General Relativity. *Science* **2018**, *360*, 1342–1346. [CrossRef] [PubMed]
43. Walsh, D.; Carswell, R.F.; Weymann, R.J. 0957+ 561 A, B: Twin quasistellar objects or gravitational lens? *Nature* **1979**, *279*, 381–384. [CrossRef]
44. Metcalf, R.B.; Meneghetti, M.; Avestruz, C.; Bellagamba, F.; Bom, C.R.; Bertin, E.; Cabanac, R.; Courbin, F.; Davies, A.; Decencière, E.; et al. The strong gravitational lens finding challenge. *Astron. Astrophys.* **2019**, *625*, A119. [CrossRef]
45. Cortes, C.; Vapnik, V. Support-vector networks. *Mach. Learn.* **1995**, *20*, 273–297. [CrossRef]

46. de Jong, J.T.; Verdoes Kleijn, G.A.; Kuijken, K.H.; Valentijn, E.A.; KiDS; Consortiums, A.W. The kilo-degree survey. *Exp. Astron.* **2013**, *35*, 25–44. [CrossRef]
47. Hartley, P.; Flamary, R.; Jackson, N.; Tagore, A.; Metcalf, R. Support vector machine classification of strong gravitational lenses. *Mon. Not. R. Astron. Soc.* **2017**, *471*, 3378–3397. [CrossRef]
48. Jacobs, C.; Glazebrook, K.; Collett, T.; More, A.; McCarthy, C. Finding strong lenses in CFHTLS using convolutional neural networks. *Mon. Not. R. Astron. Soc.* **2017**, *471*, 167–181. [CrossRef]
49. Lieu, M.; Conversi, L.; Altieri, B.; Carry, B. Detecting Solar system objects with convolutional neural networks. *Mon. Not. R. Astron. Soc.* **2019**, *485*, 5831–5842. [CrossRef]
50. Huang, X.; Baltasar, S.; Ratier-Werbin, N. DESI Strong Lens Foundry I: HST Observations and Modeling with GIGA-Lens. *arXiv* **2025**, arXiv:2502.03455.
51. Lines, N.; Collett, T.; Walmsley, M.; Rojas, K.; Li, T.; Leuzzi, L.; Manjón-García, A.; Vincken, S.; Wilde, J.; Holloway, P.; et al. Euclid Quick Data Release (Q1). The Strong Lensing Discovery Engine C-Finding lenses with machine learning. *arXiv* **2025**, arXiv:2503.15326.
52. Wilde, J.; Serjeant, S.; Bromley, J.M.; Dickinson, H.; Koopmans, L.V.; Metcalf, R.B. Detecting gravitational lenses using machine learning: Exploring interpretability and sensitivity to rare lensing configurations. *Mon. Not. R. Astron. Soc.* **2022**, *512*, 3464–3479. [CrossRef]
53. Applebaum, K.; Zhang, D. Classifying galaxy images through support vector machines. In Proceedings of the 2015 IEEE International Conference on Information Reuse and Integration, San Francisco, CA, USA, 13–15 August 2015; pp. 357–363.
54. Reza, M. Galaxy morphology classification using automated machine learning. *Astron. Comput.* **2021**, *37*, 100492. [CrossRef]
55. Dieleman, S.; Willett, K.W.; Dambre, J. Rotation-invariant convolutional neural networks for galaxy morphology prediction. *Mon. Not. R. Astron. Soc.* **2015**, *450*, 1441–1459. [CrossRef]
56. Domínguez Sánchez, H.; Huertas-Company, M.; Bernardi, M.; Tuccillo, D.; Fischer, J. Improving galaxy morphologies for SDSS with Deep Learning. *Mon. Not. R. Astron. Soc.* **2018**, *476*, 3661–3676. [CrossRef]
57. Bhambra, P.; Joachimi, B.; Lahav, O. Explaining deep learning of galaxy morphology with saliency mapping. *Mon. Not. R. Astron. Soc.* **2022**, *511*, 5032–5041. [CrossRef]
58. Muthukrishna, D.; Narayan, G.; Mandel, K.S.; Biswas, R.; Hložek, R. RAPID: Early classification of explosive transients using deep learning. *Publ. Astron. Soc. Pac.* **2019**, *131*, 118002. [CrossRef]
59. Villar, V.A.; Cranmer, M.; Berger, E.; Contardo, G.; Ho, S.; Hosseinzadeh, G.; Lin, J.Y.Y. A deep-learning approach for live anomaly detection of extragalactic transients. *Astrophys. J. Suppl. Ser.* **2021**, *255*, 24. [CrossRef]
60. Denker, J.; LeCun, Y. Transforming Neural-Net Output Levels to Probability Distributions. In *Proceedings of the Advances in Neural Information Processing Systems*; Lippmann, R., Moody, J., Touretzky, D., Eds.; Morgan-Kaufmann: San Francisco, CA, USA, 1990; Volume 3.
61. Ishida, E.E. Machine learning and the future of supernova cosmology. *Nat. Astron.* **2019**, *3*, 680–682. [CrossRef]
62. Lochner, M.; McEwen, J.D.; Peiris, H.V.; Lahav, O.; Winter, M.K. Photometric supernova classification with machine learning. *Astrophys. J. Suppl. Ser.* **2016**, *225*, 31. [CrossRef]
63. Zhang, C.; Wang, C.; Hobbs, G.; Russell, C.; Li, D.; Zhang, S.B.; Dai, S.; Wu, J.W.; Pan, Z.C.; Zhu, W.W.; et al. Applying saliency-map analysis in searches for pulsars and fast radio bursts. *Astron. Astrophys.* **2020**, *642*, A26. [CrossRef]
64. Kravtsov, A.V.; Borgani, S. Formation of galaxy clusters. *Annu. Rev. Astron. Astrophys.* **2012**, *50*, 353–409. [CrossRef]
65. Ntampaka, M.; ZuHone, J.; Eisenstein, D.; Nagai, D.; Vikhlinin, A.; Hernquist, L.; Marinacci, F.; Nelson, D.; Pakmor, R.; Pillepich, A.; et al. A deep learning approach to galaxy cluster x-ray masses. *Astrophys. J.* **2019**, *876*, 82. [CrossRef]
66. Mordvintsev, A.; Olah, C.; Tyka, M. Deepdream-a code example for visualizing neural networks. *Google Res.* **2015**, *2*, 67.
67. Collaboration, G.; Prusti, T.; de Bruijne, J.; Brown, A.; Vallenari, A.; Babusiaux, C.; Bailer-Jones, C.; Bastian, U.; Biermann, M.; Evans, D.; et al. The Gaia mission. *Astron. Astrophys.* **2016**, *595*, A1. [CrossRef]
68. Marchetti, T.; Rossi, E.; Kordopatis, G.; Brown, A.; Rimoldi, A.; Starkenburg, E.; Youakim, K.; Ashley, R. An artificial neural network to discover hypervelocity stars: Candidates in Gaia DR1/TGAS. *Mon. Not. R. Astron. Soc.* **2017**, *470*, 1388–1403. [CrossRef]
69. Van Groeningen, M.; Castro-Ginard, A.; Brown, A.; Casamiquela, L.; Jordi, C. A machine-learning-based tool for open cluster membership determination in Gaia DR3. *Astron. Astrophys.* **2023**, *675*, A68. [CrossRef]
70. Fotopoulou, S. A review of unsupervised learning in astronomy. *Astron. Comput.* **2024**, *48*, 100851. [CrossRef]
71. Malhan, K.; Ibata, R.A. STREAMFINDER—I. A new algorithm for detecting stellar streams. *Mon. Not. R. Astron. Soc.* **2018**, *477*, 4063–4076. [CrossRef]
72. Koppelman, H.H.; Helmi, A.; Massari, D.; Price-Whelan, A.M.; Starkenburg, T.K. Multiple retrograde substructures in the Galactic halo: A shattered view of Galactic history. *Astron. Astrophys.* **2019**, *631*, L9. [CrossRef]
73. Müller, A.C.; Guido, S. *Introduction to Machine Learning with Python: A Guide for Data Scientists*; O'Reilly Media, Inc.: Sebastopol, CA, USA, 2016.

74. Hoyle, B.; Rau, M.M.; Zitlau, R.; Seitz, S.; Weller, J. Feature importance for machine learning redshifts applied to SDSS galaxies. *Mon. Not. R. Astron. Soc.* **2015**, *449*, 1275–1283. [CrossRef]
75. Ribeiro, F.; Gradvohl, A.L.S. Machine learning techniques applied to solar flares forecasting. *Astron. Comput.* **2021**, *35*, 100468. [CrossRef]
76. Chaini, S.; Mahabal, A.; Kembhavi, A.; Bianco, F.B. Light curve classification with DistClassiPy: A new distance-based classifier. *Astron. Comput.* **2024**, *48*, 100850. [CrossRef]
77. Lucie-Smith, L.; Peiris, H.V.; Pontzen, A. An interpretable machine-learning framework for dark matter halo formation. *Mon. Not. R. Astron. Soc.* **2019**, *490*, 331–342. [CrossRef]
78. Davies, A.; Serjeant, S.; Bromley, J.M. Using convolutional neural networks to identify gravitational lenses in astronomical images. *Mon. Not. R. Astron. Soc.* **2019**, *487*, 5263–5271. [CrossRef]
79. Kosiba, M.; Lieu, M.; Altieri, B.; Clerc, N.; Faccioli, L.; Kendrew, S.; Valtchanov, I.; Sadibekova, T.; Pierre, M.; Hroch, F.; et al. Multiwavelength classification of X-ray selected galaxy cluster candidates using convolutional neural networks. *Mon. Not. R. Astron. Soc.* **2020**, *496*, 4141–4153. [CrossRef]
80. Jerse, G.; Marcucci, A. Deep Learning LSTM-based approaches for 10.7 cm solar radio flux forecasting up to 45-days. *Astron. Comput.* **2024**, *46*, 100786. [CrossRef]
81. Simonyan, K.; Vedaldi, A.; Zisserman, A. Deep inside convolutional networks: Visualising image classification models and saliency maps. *arXiv* **2013**, arXiv:1312.6034.
82. Sundararajan, M.; Taly, A.; Yan, Q. Axiomatic attribution for deep networks. In Proceedings of the International Conference on Machine Learning, PMLR, Sydney, Australia, 6–11 August 2017; pp. 3319–3328.
83. Selvaraju, R.R.; Cogswell, M.; Das, A.; Vedantam, R.; Parikh, D.; Batra, D. Grad-cam: Visual explanations from deep networks via gradient-based localization. In Proceedings of the IEEE International Conference on Computer Vision, Venice, Italy, 22–29 October 2017; pp. 618–626.
84. Walmsley, M.; Lintott, C.; Géron, T.; Kruk, S.; Krawczyk, C.; Willett, K.W.; Bamford, S.; Kelvin, L.S.; Fortson, L.; Gal, Y.; et al. Galaxy Zoo DECaLS: Detailed visual morphology measurements from volunteers and deep learning for 314 000 galaxies. *Mon. Not. R. Astron. Soc.* **2022**, *509*, 3966–3988. [CrossRef]
85. Jacobs, C.; Glazebrook, K.; Qin, A.K.; Collett, T. Exploring the interpretability of deep neural networks used for gravitational lens finding with a sensitivity probe. *Astron. Comput.* **2022**, *38*, 100535. [CrossRef]
86. Springenberg, J.T.; Dosovitskiy, A.; Brox, T.; Riedmiller, M. Striving for simplicity: The all convolutional net. *arXiv* **2014**, arXiv:1412.6806.
87. Selvaraju, R.R.; Das, A.; Vedantam, R.; Cogswell, M.; Parikh, D.; Batra, D. Grad-CAM: Why did you say that? *arXiv* **2016**, arXiv:1611.07450.
88. Tanoglidis, D.; Ćiprijanović, A.; Drlica-Wagner, A. DeepShadows: Separating low surface brightness galaxies from artifacts using deep learning. *Astron. Comput.* **2021**, *35*, 100469. [CrossRef]
89. Shirasuna, V.Y.; Gradvohl, A.L.S. An optimized training approach for meteor detection with an attention mechanism to improve robustness on limited data. *Astron. Comput.* **2023**, *45*, 100753. [CrossRef]
90. Adebayo, J.; Gilmer, J.; Muelly, M.; Goodfellow, I.; Hardt, M.; Kim, B. Sanity checks for saliency maps. In Proceedings of the Advances in Neural Information Processing Systems 31 (NeurIPS 2018), Montréal, QC, Canada, 3–8 December 2018.
91. Srinivas, S.; Fleuret, F. Rethinking the role of gradient-based attribution methods for model interpretability. *arXiv* **2020**, arXiv:2006.09128.
92. Shapley, L.S. Stochastic games. *Proc. Natl. Acad. Sci. USA* **1953**, *39*, 1095–1100. [CrossRef]
93. Heyl, J.; Butterworth, J.; Viti, S. Understanding molecular abundances in star-forming regions using interpretable machine learning. *Mon. Not. R. Astron. Soc.* **2023**, *526*, 404–422. [CrossRef]
94. Ye, S.; Cui, W.Y.; Li, Y.B.; Luo, A.L.; Jones, R.H. Deep learning interpretable analysis for carbon star identification in Gaia DR3. *Astron. Astrophys.* **2024**, *697*, A107. [CrossRef]
95. Goh, K.M.; Lim, D.H.Y.; Sham, Z.D.; Prakash, K.B. An Interpretable Galaxy Morphology Classification Approach using Modified SqueezeNet and Local Interpretable Model-Agnostic Explanation. *Res. Astron. Astrophys.* **2025**, *25*, 065018. [CrossRef]
96. Breiman, L.; Friedman, J.; Olshen, R.A.; Stone, C.J. *Classification and Regression Trees*; Routledge: London, UK, 2017.
97. Friedman, J.H.; Popescu, B.E. Predictive Learning via Rule Ensembles. *Ann. Appl. Stat.* **2008**, *2*, 916–954. [CrossRef]
98. Odewahn, S.; De Carvalho, R.; Gal, R.; Djorgovski, S.; Brunner, R.; Mahabal, A.; Lopes, P.; Moreira, J.K.; Stalder, B. The Digitized Second Palomar Observatory Sky Survey (DPOSS). III. Star-Galaxy Separation. *Astron. J.* **2004**, *128*, 3092. [CrossRef]
99. Bailey, S.; Aragon, C.; Romano, R.; Thomas, R.C.; Weaver, B.A.; Wong, D. How to find more supernovae with less work: Object classification techniques for difference imaging. *Astrophys. J.* **2007**, *665*, 1246. [CrossRef]
100. Chacón, J.; Vázquez, J.A.; Almaraz, E. Classification algorithms applied to structure formation simulations. *Astron. Comput.* **2022**, *38*, 100527. [CrossRef]

101. Bahdanau, D.; Cho, K.; Bengio, Y. Neural Machine Translation by Jointly Learning to Align and Translate. *arXiv* **2016**, arXiv:1409.0473.
102. Jain, S.; Wallace, B.C. Attention is not explanation. *arXiv* **2019**, arXiv:1902.10186.
103. Wiegreffe, S.; Pinter, Y. Attention is not not explanation. *arXiv* **2019**, arXiv:1908.04626.
104. Bowles, M.; Scaife, A.M.; Porter, F.; Tang, H.; Bastien, D.J. Attention-gating for improved radio galaxy classification. *Mon. Not. R. Astron. Soc.* **2021**, *501*, 4579–4595. [CrossRef]
105. Koza, J.R. Genetic programming as a means for programming computers by natural selection. *Stat. Comput.* **1994**, *4*, 87–112. [CrossRef]
106. Schmidt, M.; Lipson, H. Distilling free-form natural laws from experimental data. *Science* **2009**, *324*, 81–85. [CrossRef] [PubMed]
107. Cranmer, M. Interpretable machine learning for science with PySR and SymbolicRegression.jl. *arXiv* **2023**, arXiv:2305.01582.
108. Udrescu, S.M.; Tegmark, M. AI Feynman: A physics-inspired method for symbolic regression. *Sci. Adv.* **2020**, *6*, eaay2631. [CrossRef] [PubMed]
109. Cranmer, M.; Sanchez Gonzalez, A.; Battaglia, P.; Xu, R.; Cranmer, K.; Spergel, D.; Ho, S. Discovering symbolic models from deep learning with inductive biases. In Proceedings of the Advances in Neural Information Processing Systems 33 (NeurIPS 2020), Virtual, 6–12 December 2020; pp. 17429–17442.
110. Jin, Z.; Davis, B.L. Discovering Black Hole Mass Scaling Relations with Symbolic Regression. *arXiv* **2023**, arXiv:2310.19406.
111. Sousa-Neto, A.; Bengaly, C.; González, J.E.; Alcaniz, J. No evidence for dynamical dark energy from DESI and SN data: A symbolic regression analysis. *arXiv* **2025**, arXiv:2502.10506.
112. Wadekar, D.; Villaescusa-Navarro, F.; Ho, S.; Perreault-Levasseur, L. Modeling assembly bias with machine learning and symbolic regression. *arXiv* **2020**, arXiv:2012.00111.
113. Guo, N.; Lucie-Smith, L.; Peiris, H.V.; Pontzen, A.; Piras, D. Deep learning insights into non-universality in the halo mass function. *Mon. Not. R. Astron. Soc.* **2024**, *532*, 4141–4156. [CrossRef]
114. Klys, J.; Snell, J.; Zemel, R. Learning latent subspaces in variational autoencoders. In Proceedings of the Advances in Neural Information Processing Systems 31 (NeurIPS 2018), Montréal, QC, Canada, 3–8 December 2018.
115. Raissi, M.; Perdikaris, P.; Karniadakis, G.E. Physics-informed neural networks: A deep learning framework for solving forward and inverse problems involving nonlinear partial differential equations. *J. Comput. Phys.* **2019**, *378*, 686–707. [CrossRef]
116. Wang, S.; Teng, Y.; Perdikaris, P. Understanding and mitigating gradient flow pathologies in physics-informed neural networks. *SIAM J. Sci. Comput.* **2021**, *43*, A3055–A3081. [CrossRef]
117. Caruana, R.; Kangarloo, H.; Dionisio, J.D.; Sinha, U.; Johnson, D. Case-based explanation of non-case-based learning methods. In Proceedings of the AMIA Symposium, Washington, DC, USA, 6–10 November 1999; p. 212.
118. Wei, J.; Wang, X.; Schuurmans, D.; Bosma, M.; Ichter, B.; Xia, F.; Chi, E.; Le, Q.V.; Zhou, D. Chain-of-thought prompting elicits reasoning in large language models. In Proceedings of the Advances in Neural Information Processing Systems 35 (NeurIPS 2022), New Orleans, LA, USA, 28 November–9 December 2022; pp. 24824–24837.
119. Yugeswardeenoo, D.; Zhu, K.; O'Brien, S. Question-analysis prompting improves LLM performance in reasoning tasks. *arXiv* **2024**, arXiv:2407.03624.
120. Turpin, M.; Michael, J.; Perez, E.; Bowman, S. Language models don't always say what they think: Unfaithful explanations in chain-of-thought prompting. In Proceedings of the Advances in Neural Information Processing Systems 36 (NeurIPS 2023), New Orleans, LA, USA, 10–16 December 2023; pp. 74952–74965.
121. Li, J.; Cao, P.; Chen, Y.; Liu, K.; Zhao, J. Towards faithful chain-of-thought: Large language models are bridging reasoners. *arXiv* **2024**, arXiv:2405.18915.
122. Arcuschin, I.; Janiak, J.; Krzyzanowski, R.; Rajamanoharan, S.; Nanda, N.; Conmy, A. Chain-of-thought reasoning in the wild is not always faithful. *arXiv* **2025**, arXiv:2503.08679.
123. Lanham, T.; Chen, A.; Radhakrishnan, A.; Steiner, B.; Denison, C.; Hernandez, D.; Li, D.; Durmus, E.; Hubinger, E.; Kernion, J.; et al. Measuring faithfulness in chain-of-thought reasoning. *arXiv* **2023**, arXiv:2307.13702.
124. Bentham, O.; Stringham, N.; Marasović, A. Chain-of-thought unfaithfulness as disguised accuracy. *arXiv* **2024**, arXiv:2402.14897.
125. Wang, Z.; Han, Z.; Chen, S.; Xue, F.; Ding, Z.; Xiao, X.; Tresp, V.; Torr, P.; Gu, J. Stop Reasoning! When Multimodal LLM with Chain-of-Thought Reasoning Meets Adversarial Image. *arXiv* **2024**, arXiv:2402.14899.
126. Yamada, Y.; Lange, R.T.; Lu, C.; Hu, S.; Lu, C.; Foerster, J.; Clune, J.; Ha, D. The AI Scientist-v2: Workshop-Level Automated Scientific Discovery via Agentic Tree Search. *arXiv* **2025**, arXiv:2504.08066.
127. Moss, A. The AI Cosmologist I: An Agentic System for Automated Data Analysis. *arXiv* **2025**, arXiv:2504.03424.
128. Fluke, C.J.; Jacobs, C. Surveying the reach and maturity of machine learning and artificial intelligence in astronomy. *WIREs Data Min. Knowl. Discov.* **2020**, *10*, e1349. [CrossRef]
129. Khakzar, A.; Baselizadeh, S.; Navab, N. Rethinking positive aggregation and propagation of gradients in gradient-based saliency methods. *arXiv* **2020**, arXiv:2012.00362.
130. Zhang, Y.; Gu, S.; Song, J.; Pan, B.; Bai, G.; Zhao, L. XAI benchmark for visual explanation. *arXiv* **2023**, arXiv:2310.08537.

131. Moiseev, I.; Balabaeva, K.; Kovalchuk, S. Open and Extensible Benchmark for Explainable Artificial Intelligence Methods. *Algorithms* **2025**, *18*, 85. [CrossRef]
132. Information Commissioner's Office. *Guidance on AI and Data Protection*; Information Commissioner's Office: Cheshire, UK, 2023. Available online:https://ico.org.uk/for-organisations/uk-gdpr-guidance-and-resources/artificial-intelligence/guidance-on-ai-and-data-protection/ (accessed on 23 April 2025).
133. Bereska, L.; Gavves, E. Mechanistic Interpretability for AI Safety—A Review. *arXiv* **2024**, arXiv:2404.14082.
134. Bluck, A.F.L.; Maiolino, R.; Brownson, S.; Conselice, C.J.; Ellison, S.L.; Piotrowska, J.M.; Thorp, M.D. The quenching of galaxies, bulges, and disks since cosmic noon. A machine learning approach for identifying causality in astronomical data. *Astron. Astrophys.* **2022**, *659*, A160. [CrossRef]
135. Moschou, S.; Hicks, E.; Parekh, R.; Mathew, D.; Majumdar, S.; Vlahakis, N. Physics-informed neural networks for modeling astrophysical shocks. *Mach. Learn. Sci. Technol.* **2023**, *4*, 035032. [CrossRef]
136. Ni, S.; Qiu, Y.; Chen, Y.; Song, Z.; Chen, H.; Jiang, X.; Quan, D.; Chen, H. Application of Physics-Informed Neural Networks in Removing Telescope Beam Effects. *arXiv* **2024**, arXiv:2409.05718.
137. Mengaldo, G. Explain the Black Box for the Sake of Science: The Scientific Method in the Era of Generative Artificial Intelligence. *arXiv* **2025**, arXiv:2406.10557.

Disclaimer/Publisher's Note: The statements, opinions and data contained in all publications are solely those of the individual author(s) and contributor(s) and not of MDPI and/or the editor(s). MDPI and/or the editor(s) disclaim responsibility for any injury to people or property resulting from any ideas, methods, instructions or products referred to in the content.

MDPI AG
Grosspeteranlage 5
4052 Basel
Switzerland
Tel.: +41 61 683 77 34

Universe Editorial Office
E-mail: universe@mdpi.com
www.mdpi.com/journal/universe

Disclaimer/Publisher's Note: The title and front matter of this reprint are at the discretion of the Guest Editors. The publisher is not responsible for their content or any associated concerns. The statements, opinions and data contained in all individual articles are solely those of the individual Editors and contributors and not of MDPI. MDPI disclaims responsibility for any injury to people or property resulting from any ideas, methods, instructions or products referred to in the content.

www.ingramcontent.com/pod-product-compliance
Lightning Source LLC
Chambersburg PA
CBHW061621210926
51850CB00009B/2395
* 9 7 8 3 7 2 5 8 5 5 6 9 8 *